ZIML Math Competition Book

Division E 2016-2017

Areteem Institute

ZIML Math Competition Book Division E 2016-2017

Edited by   John Lensmire
David Reynoso
Kevin Wang
Kelly Ren

Cover and chapter title photographs by Kelly Ren and Kevin Wang

First printing, April 2018.

Contents

Introduction

Each month during the school year, Areteem Institute hosts the online Zoom International Math League (ZIML) competitions. Students can compete in one of five divisions based on their age and mathematical level.

The book contains the problems, answers, and full solutions from the nine ZIML Division E Competitions held during the 2016-2017 School Year. It is divided into three parts:

1. The complete Division E ZIML Competitions (20 questions per competition) from October 2016 to June 2017.
2. The solutions for each of the competitions, including detailed work and helpful tricks.
3. An appendix including the topics and knowledge points covered for Division E, a glossary including common mathematical terms, and answer keys for each of the competitions so students can easily check their work.

The questions found on the ZIML competitions are meant to test your problem solving skills and train you to apply the knowledge you know to many different applications. We hope you enjoy the problems!

About Zoom International Math League

The Zoom International Math League (ZIML) has a simple goal: provide a platform for students to build and share their passion for math and other STEM fields with students from around the globe. Started in 2008 as the Southern California Mathematical Olympiad, ZIML has a rich history of past participants who have advanced to top tier colleges and prestigious math competitions, including American Math Competitions, MATHCOUNTS, and the International Math Olympaid.

The ZIML Core Online Programs, most available with a free account at ziml.areteem.org, include:

- **Daily Magic Spells:** Provides a problem a day (Monday through Friday) for students to practice, with full solutions available the next day.
- **Weekly Brain Potions:** Provides one problem per week posted in the online discussion forum at ziml.areteem.org. Usually the problem does not have a simple answer, and students can join the discussion to share their thoughts regarding the scenarios described in the problem, explore the math concepts behind the problem, give solutions, and also ask further questions.
- **Monthly Contests:** The ZIML Monthly Contests are held the first weekend of each month during the school year (October through June). Students can compete in one of 5 divisions to test their knowledge and determine their strengths and weaknesses, with winners announced after the competition.
- **Math Competition Practice:** The Practice page contains sample ZIML contests and an archive of AMC-series tests for online practice. The practices simulate the real contest environment with time-limits of the contests automatically controlled by the server.
- **Online Discussion Forum:** The Online Discussion Forum

is open for any comments and questions. Other discussions, such as hard Daily Magic Spells or the Weekly Brain Potions are also posted here.

These programs encourage students to participate consistently, so they can track their progress and improvement each year.

In addition to the online programs, ZIML also hosts onsite Local Tournaments and Workshops in various locations in the United States. Each summer, there are onsite ZIML Competitions at held at Areteem Summer Programs, including the National ZIML Convention, which is a two day convention with one day of workshops and one day of competition.

ZIML Monthly Contests are organized into five divisions ranging from upper elementary school to advanced material based on high school math.

- **Varsity:** This is the top division. It covers material on the level of the last 10 questions on the AMC 12 and AIME level. This division is open to all age levels.
- **Junior Varsity:** This is the second highest competition division. It covers material at the AMC 10/12 level and State and National MathCounts level. This division is open to all age levels.
- **Division H:** This division focuses on material from a standard high school curriculum. It covers topics up to and including pre-calculus. This division will serve as excellent practice for students preparing for the math portions of the SAT or ACT. This division is open to all age levels.
- **Division M:** This division focuses on problem solving using math concepts from a standard middle school math curriculum. It covers material at the level of AMC 8 and School or Chapter MathCounts. This division is open to all students who have not started grade 9.

- **Division E:** This division focuses on advanced problem solving with mathematical concepts from upper elementary school. It covers material at a level comparable to MOEMS Division E. This division is open to all students who have not started grade 6.

This problem book features the Division E Contests. For a detailed list of topics covered for Division E see p.149 in the Appendix.

About Areteem Institute

Areteem Institute is an educational institution that develops and provides in-depth and advanced math and science programs for K-12 (Elementary School, Middle School, and High School) students and teachers. Areteem programs are accredited supplementary programs by the Western Association of Schools and Colleges (WASC). Students may attend the Areteem Institute through these options:

- Live and real-time face-to-face online classes with audio, video, interactive online whiteboard, and text chatting capabilities;
- Self-paced classes by watching the recordings of the live classes;
- Short video courses for trending math, science, technology, engineering, English, and social studies topics;
- Summer Intensive Camps on prestigious university campuses and Winter Boot Camps;
- Practice with selected daily problems for free, and monthly ZIML competitions at ziml.areteem.org.

The Areteem courses are designed and developed by educational experts and industry professionals to bring real world applications into STEM education. The programs are ideal for students who wish to build their mathematical strength in order to excel academically and eventually win in Math Competitions (AMC, AIME, USAMO, IMO, ARML, MathCounts, Math Olympiad, ZIML, and other math leagues and tournaments, etc.), Science Fairs (County Science Fairs, State Science Fairs, national programs like Intel Science and Engineering Fair, etc.) and Science Olympiad, or purely want to enrich their academic lives by taking more challenges and developing outstanding analytical, logical thinking and creative problem solving skills.

Since 2004 Areteem Institute has been teaching with methodology that is highly promoted by the new Common Core State Standards: stressing the conceptual level understanding of the math concepts, problem solving techniques, and solving problems with real world applications. With the guidance from experienced and passionate professors, students are motivated to explore concepts deeper by identifying an interesting problem, researching it, analyzing it, and using a critical thinking approach to come up with multiple solutions.

Thousands of math students who have been trained at Areteem achieved top honors and earned top awards in major national and international math competitions, including Gold Medalists in the International Math Olympiad (IMO), top winners and qualifiers at the USA Math Olympiad (USAMO/JMO), and AIME, top winners at the Zoom International Math League (ZIML), and top winners at the MathCounts National. Many Areteem Alumni have graduated from high school and gone on to enter their dream colleges such as MIT, Cal Tech, Harvard, Stanford, Yale, Princeton, U Penn, Harvey Mudd College, UC Berkeley, UCLA, etc. Those who have graduated from colleges are now playing important roles in their fields of endeavor.

Further information about Areteem Institute, as well as updates and errata of this book, can be found online at http://www.areteem.org.

Acknowledgments

This book contains the Online ZIML Division E Problems from the 2016-17 school year. These problems were created and compiled by the staff of Areteem Institute. These problems were inspired by questions from the Areteem Math Challenge Courses, past questions on the ACT/SAT/GRE, past math competitions, math textbooks, and countless other resources and people encountered by the Areteem Curriculum Department in their life devoted to math. We thank all these sources for growing and nurturing our passion for math.

The Areteem staff, including John Lensmire, David Reynoso, Kevin Wang, and Kelly Ren, are the main contributors who compiled, edited, and reviewed this book. Photographs included on the cover and chapter introduction pages are credit to Kelly Ren and Kevin Wang.

Lastly, thanks to all the students who have participated and continue to participate in the Zoom International Math League. Your dedication to the Daily Magic Spells and Monthly Contests makes all of this possible, and we hope you continue to enjoy ZIML for years to come!

1. ZIML Contests

This part of the book contains the Division E ZIML Contests from the 2016-17 School Year. There were nine monthly competitions, held on the dates found below:

- October 7-8
- November 4-6
- December 2-4
- January 6-8
- February 3-5
- March 3-5
- April 7-9
- May 5-7
- June 2-4

1.1 ZIML October 2016 Division E

Below are the 20 Problems from the Division E ZIML Competition held in October 2016.
The answer key is available on p.158 in the Appendix.
Full solutions to these questions are available starting on p.82.

Problem 1
Calculate $42 + 33 + 17 + 58$.

Problem 2
There are 3 cows and 4 horses on the farm. Each cow eats 16 bundles of hay per day, while each horse eats 14 bundles of hay each day (the farmer makes small bundles). How many bundles of hay must the farmer keep ready each week?

Problem 3
Suppose you write the decimal 0.48 as a fraction in lowest terms with numerator N and denominator M. What is M?

Problem 4

In the diagram, points A, B, C, D are in the middle of the sides of the rectangle.

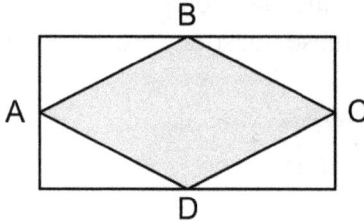

Suppose the area of the whole rectangle is 30. What is the area of the shaded region?

Problem 5

If the tens digit of a number is 2, the ones digit is twice the tens digit, and the hundreds digit is twice the ones digit, then what is the number?

Problem 6

A circle and two distinct lines are drawn on a sheet of paper. What is the largest possible number of points of intersection of these figures?

Problem 7

Josh goes shopping for socks. Each pair of socks costs $8. Josh buys as many socks as he can with $100. If Josh pays for these socks with a $100 dollar bill, how many dollars is his change?

Problem 8
Determine the 20th number in this sequence: 4, 0, −4, −8, −12,...

Problem 9
At your school party, a big basket of sandwiches was delivered that weighs a total of 19 lbs. After half of the sandwiches were taken, the basket with the left-over sandwiches weighs 12 lbs. How much does the basket weigh in lbs?

Problem 10
If 8 people are in a room and everyone shakes everyone else's hand once, then how many handshakes were in the room?

Problem 11
Consider the two squares in the diagram below.

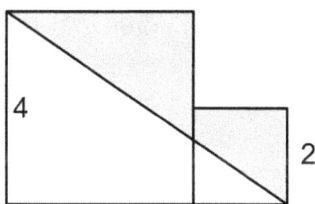

What is the area of the shaded region?

Problem 12
Suppose two trains leave from the same station at the same time, but move in opposite directions. One train averages 56 miles per hour and the other averages 64 miles per hour. How far apart in miles will the trains be at the end of three hours?

Problem 13
Consider the pattern $4, 3, 3, 4, 6, 9, 13, 18, \ldots$. What is the next number in the pattern?

Problem 14
A school bought some basketballs and volleyballs, 50 balls in total. There are 10 more basketballs than volleyballs. How many volleyballs are there?

Problem 15
Jamie counted the number of edges of a cube, Jimmy counted the number of corners, and Judy counted the number of faces. They then added the three numbers. What was the resulting sum?

Problem 16
The number 123456789 is repeatedly written producing the pattern 123456789123456879123456789.... If the pattern is continued, what number will occur in the 1009th position?

Problem 17
A book is opened, and the product of the two facing page numbers that appear is 1056. What is the smaller of the two page numbers?

Problem 18
Alice, Beth, and Cynthia have 50 dollars in total. Alice has twice as much money as Beth, and Beth has 3 times as much as Cynthia. How many dollars does Beth have?

Problem 19
In the diagram below, the area of the large circle is 100. Find the area of the shaded region.

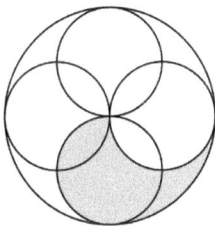

Problem 20
On a twenty-question test, each correct answer is worth 5 points, each unanswered question is worth 1 point and each incorrect answer is worth 0 points. How many different scores greater than or equal to 90 are possible?

1.2 ZIML November 2016 Division E

Below are the 20 Problems from the Division E ZIML Competition held in November 2016.
The answer key is available on p.159 in the Appendix.
Full solutions to these questions are available starting on p.87.

Problem 1

Lucas was bored and decided to use his favorite highlighter to highlight words on his brother's favorite Harry Potter book. He spent 2 seconds to highlight each word, and before his brother could stop him, he already highlighted 450 words. How many minutes in total did it take him to highlight those words?

Problem 2

Mike and Scott each have candy bars. Mike's candy bar is 3 inches wide and 8 inches long. Scott's is 3 inches wide and 6 inches long. Scott eats half of his candy bar. What percent of his candy bar should Mike eat so that he will eat the same amount of candy as Scott? Round your answer to the nearest tenth. For example, input 33.33% as 33.3.

Problem 3

Consider the pattern starting with

$$5, 7, 6, 8, 7, 9, 8, \ldots$$

What is the 20th number in the pattern?

Problem 4

Suppose the fraction

$$\frac{1}{2} \times \frac{2}{3} \times \frac{3}{4} \times \frac{4}{5} \times \frac{5}{6} \times \frac{6}{7} \times \frac{7}{8} \times \frac{8}{9} \times \frac{9}{10}$$

is simplified and written with numerator N and denominator M. What is $N + M$?

Problem 5

Little Pig got 24 bricks and built a wall along one side of his garden. The wall was 4 bricks high and 5 bricks across. Little Pig painted all of the exposed faces of the bricks.

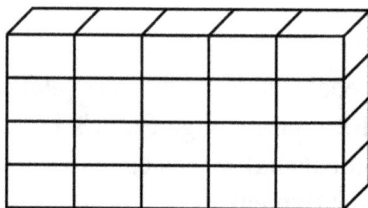

How many faces of bricks did Little Pig have to paint?

Problem 6

What is the value of the sum $-5 + 10 - 15 + 20 - 25 + 30 - 35 + 40 - 45 \cdots + 100$?

Problem 7

Consider the diagram of a pentagon below (with 5 equal sides and 5 equal angles).

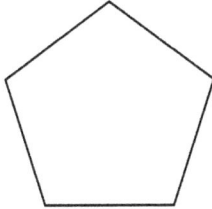

How many lines of symmetry does the pentagon have?

Problem 8

Suppose you have a lot of $\$1, \$2, \$4, \$7$ bills. What is the fewest number of total bills needed in order to make $\$41$ exactly?

Problem 9

David and Ed went to pick cherries. Together they picked 82 cherries. If David gave 4 cherries to Ed, they would have the same number of cherries. How many cherries did David pick?

Problem 10

How many different four-digit numbers can be formed by rearranging the four digits in 3001? Note: 0 cannot be the leading digit, so 0031 does not count, but 3001 itself does.

Problem 11

A rectangle has perimeter 18. If both the length and the width of the rectangle are whole numbers, what is the largest possible area for the rectangle?

Problem 12

The first few terms of a pattern are shown below.

$$\leftarrow, \rightarrow, \uparrow, \downarrow, \rightarrow, \leftarrow, \downarrow, \uparrow, \leftarrow, \rightarrow, \uparrow, \ldots.$$

In the first 18 terms, how many times is a left arrow immediately followed by a right arrow?

Problem 13

When Matt was 17 years old, Nathan was 23. This year the sum of their ages is 50. How old is Matt this year?

Problem 14

In a mathematics contest with ten problems, a student gains 5 points for a correct answer and loses 2 points for an incorrect answer. If Olivia answered every problem and her score was 29, how many correct answers did she have?

Problem 15

Ten cubic feet hold about 75 gallons of water. William has a pool that is 10 feet wide, 15 feet long, and 6 feet deep. How many gallons of water does William need to fill his pool?

Problem 16

The number 3113 is a palindrome (a number that reads the same from left to right as it does from right to left). What is the product of the digits of the smallest number larger than 3113 that is a palindrome?

Problem 17

A square with area 9 can be divided into 9 unit squares. In total, this creates squares with three different sizes, including 9 unit squares, 4 squares with area 4, and the 1 original square with area 9, or $9 + 4 + 1 = 14$ squares in total. How many squares in total are created when a square with area 25 is divided into unit squares?

Problem 18

A book has 100 pages numbered 1, 2, 3, and so on. How many times does the digit 1 appear in the page numbers?

Problem 19

A five-legged Martian has a drawer full of socks, each of which is red, white, or blue, and there are at least five socks of each color. The Martian pulls out one sock at a time without looking. How many socks must the Martian take from the drawer to be certain there will be 5 socks of the same color?

Problem 20

Consider the diagram below:

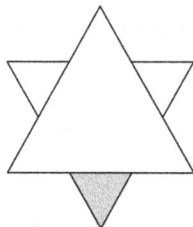

The big triangle has three equal sides. To draw the three identical small triangles, each side of the big triangle was divided into three equal pieces, and the middle piece was used as the side of the smaller triangles. If each of the small triangles have three equal sides and the shaded small triangle below has area 4, what is the are of the big triangle?

1.3   ZIML December 2016 Division E

Below are the 20 Problems from the Division E ZIML Competition held in December 2016.
The answer key is available on p.160 in the Appendix.
Full solutions to these questions are available starting on p.93.

Problem 1
Calculate $25 \times 3 \times 7 \times 4$.

Problem 2
A number has 4 digits. It's ones digit is the second largest one digit number. The tens digit is half the ones digit. The hundreds digit is one more than half of the tens digit. The thousands digit is the difference between the hundreds and tens digit. What is the number?

Problem 3
Consider a square drawn inside another square, creating 4 identical triangles as shown below.

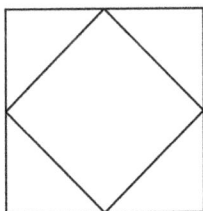

If the inner square has an area of 20 cm^2, what is the area of ONE of the triangles measured in cm^2?

Problem 4

Melissa drove for 90 minutes at a rate of 50 miles per hour and for 30 minutes at 60 miles per hour. How many miles did she travel in total?

Problem 5

Three identical rectangles combine to make a large rectangle as shown below.

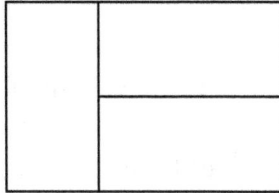

If the longer side of each of the three smaller rectangles is 4 inches, what is the area of the large rectangle in square inches?

Problem 6

When gold sold for $16 an ounce, Johnny found $48 worth of gold in his claim. Gold presently sells for $300 an ounce. How many dollars is Johnny's amount of gold worth today?

Problem 7

Grace and Andrew enjoy building towers with blocks. Grace has blocks that are 10 cm high and Andrew has blocks that are 6 cm high. They want to both build towers that have the same height. What is the smallest height (in cm) that both Grace and Andrew can build with their blocks?

ZOOM INTERNATIONAL MATH LEAGUE: ziml.areteem.org

Problem 8
Amy and her sister Claire's house has a 420 square foot lawn in the back. Amy can mow 120 square feet in 30 minutes. When Amy and Claire work together, they can finish the whole lawn in one hour. How many square feet per minute can Claire mow?

Problem 9
Six trees are equally spaced along one side of a straight road. The distance from the first tree to the fourth is 60 feet. What is the distance in feet between the first and last trees?

Problem 10
In

$$2 \times \bigstar + 3 \times \bigstar + 4 \times \bigstar + 5 \times \bigstar + 6 \times \bigstar + 7 \times \bigstar = 2700,$$

$\bigstar$ represents the same number. Which number is $\bigstar$?

Problem 11
Some cars and motorcycles were parked in a parking lot near William's house. (Cars have 4 wheels, and motorcycles have 2 wheels.) William counted the cars and motorcycles together, 7 in total, and there were 18 wheels. How many cars were there?

Problem 12

Drawing two circles, it is possible to divide the circles into 3 regions, as in the diagram below.

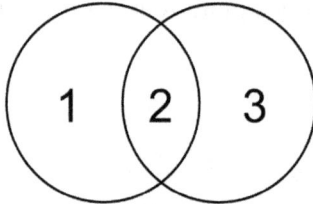

If we draw add third circle to divide the circles into as many regions as possible, how many regions do we get?

Problem 13

Harry Happenstance happened upon a pride of lions sitting underneath a tree. The leaves on the tree had an interesting property. On the first branch there was one leaf, on the second branch there was also one leaf, but on the third branch there were two leaves. From then on the number of leaves on the next branch was the sum of the number of leaves on the two branches below it. How many leaves were on the tenth branch?

Problem 14

Suppose you have a box (rectangular prism) that is 2 feet long, 1 foot wide, and has a height of 6 inches. You want to store 2 inch by 2 inch by 2 inch cubes in the box. How many of these cubes can you store in the box?

Problem 15
We planted 121 different plants in the school's garden. Unfortunately some of the plants got eaten by squirrels. The number of plants that survived is 61 more than the number of plants that were eaten by the squirrels. How many of the plants did the squirrels eat?

Problem 16
Bill and Will are drawing triangles. Bill drew the right triangle shown on the left side below. Will tried to draw the same triangle, but ended up drawing his triangle too big. Will's similar triangle is shown on the right.

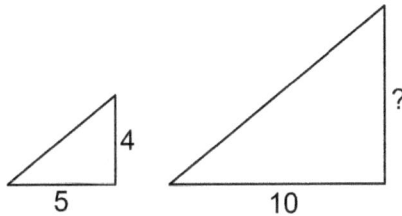

What is the missing height of Will's triangle?

Problem 17
Find the greatest 2-digit number that leaves a remainder of 4 when divided by 11.

Problem 18

Consider the square *ABCD* divided into 16 small squares as in the image below.

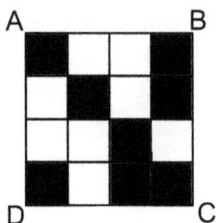

In this case, 4 small squares in the top half of *ABCD* are shaded black. Suppose the image is rotated so that *AD* is on the bottom. After this rotation, how many small squares in the top half are shaded black?

Problem 19

I just found a box with Christmas ornaments in the attic. It has a huge label on the front that says it contains 200 Christmas ornaments of three different colors (red, blue, and white). I'm so clumsy that I broke some of them by accident. It seems I broke 14 white ornaments and now I have the same number of white ornaments and blue ornaments. Mom says that before I broke them we had 4 more red ornaments than 3 times the number of white ornaments. How many red ornaments were there in the box before I found it?

Problem 20

Brandon and Cathy is taking their kids, Alex and David, on a road trip. Their car has 4 seats: one driver's seat and one passenger seat in the front, and two back seats. Both Brandon and Cathy can drive, but the kids are too young to drive. If Alex always sits next to Brandon, how many possible seating arrangements are there?

1.4   ZIML January 2017 Division E

Below are the 20 Problems from the Division E ZIML Competition held in January 2017.
The answer key is available on p.161 in the Appendix.
Full solutions to these questions are available starting on p.101.

Problem 1
Calculate the following sum:

$$88 + 86 + 91 + 92 + 87 + 90 + 89 + 93 + 92 + 88$$

Problem 2
Suppose the height of an elf is 3 times the height of a hobbit. Also suppose that an elf is 60 inches taller than a hobbit. What is the height of the elf in inches?

Problem 3
How many lines of symmetry does the pinwheel shape below have?

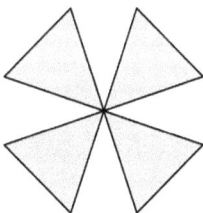

Problem 4

Timmy is getting ready to go to a party and is having trouble deciding what to wear. He has 3 different pairs of jeans, 8 different shirts, and 2 different pairs of shoes. In how many different ways can he get dressed?

Problem 5

George's family lives in a house with a four-digit street number. The difference of the first digit and the last digit is 8. The 2nd digit is twice the first digit, and the 3rd digit is twice the 2nd digit. What is the street number of George's house?

Problem 6

David gives out fruit snacks as awards for the students in his math class for solving problems and presenting solutions in front of the class. He starts with a full bag of fruit snacks. Each day, he gives out 25% of the fruit snacks in the bag at the beginning of that day. At the end of the second day, 36 fruit snacks remain in the bag. How many fruit snacks were in the bag originally?

Problem 7

Find the area of the shaded region below where the diagram is made up of a semicircle and a square.

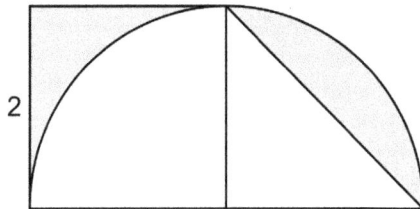

Problem 8

Captain Hook, Popeye, and Sinbad went out to the sea to hunt for treasure. They all found diamonds. The total number of diamonds found by Captain Hook and Popeye is 80. The total number of diamonds found by Popeye and Sinbad is 70. The total number of diamonds found by Captain Hook and Sinbad is 50. How many diamonds did Captain Hook find?

Problem 9

Your cousin asked you for help setting up the numbers for the tables at her wedding. She gave you lots of stickers with all the available digits: $0, 1, 2, \ldots, 9$. You need to label tables starting with 1 and ending with 150. How many of the digit 2 stickers will you use?

Problem 10

One pipe can fill a swimming pool 1.5 times as fast as a second pipe. If the gardener opens both pipes, they fill the pool in 10 hours. How many hours would it take to fill the pool if only the slower pipe is used?

Problem 11

Start with a 12×12 square. Shade in a 4×4 square in the center, and shade 4 triangles with base and height 4 in the corners as shown in the diagram below.

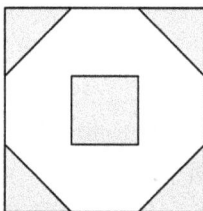

What is the area of the unshaded portion of the square?

Problem 12

Suppose a truck travels in segments that are described in the table below:

Segment	Distance (miles)	Time (hours)
1	30	1
2	90	2
3	50	1

What is the average speed of the truck in miles per hour? Express your answer as a decimal rounded to the nearest tenth.

Problem 13

William's family loves to come visit him in California. His uncle comes to visit every 2 months and his grandmother comes to visit every 3 months. If both his uncle and grandmother visited him in December 2016, how many months will there be in 2017 where William is NOT visited by either his uncle or grandmother?

Problem 14

If a 4×2 rectangle is rolled once on its side it moves a total of 4 units as shown in the diagram below.

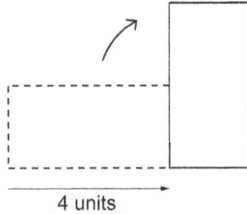

4 units

If a 5×3 rectangle is rolled 10 times on its side, how many units does it move?

Problem 15

In the diagram, each side is perpendicular to its adjacent sides, and all small sides have equal length. Given that the perimeter of this diagram is 108cm, find the area of the shape.

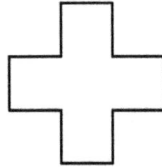

Problem 16

Thomas and Thelma go to the zoo to look at the exhibit containing lions and tigers. The brochure states that there are exactly 11 animals in the exhibit. Thomas shows Thelma the 5 tigers that he sees in the exhibit. After looking for herself, Thelma says she sees twice as many tigers as lions. If all the above statements are correct (but Thomas and Thelma do not necessarily see all the animals in the exhibit), what is the maximum number of lions in the exhibit?

Problem 17

A convenience store owner wishes to mix together raisins and roasted peanuts to produce a high energy snack for hikers. The raisins sell for $3.50 per kilogram and the nuts sell for $4.75 per kilogram. 20 kg of the snack are made with a price of $4.00 per kilogram. How many kg of this mixture is raisins?

Problem 18

Sophia builds a chest out of wood (in the shape of a box). The outside of the box is 8 inches long, 5 inches wide, and 4 inches tall. The chest is 1 inch thick on all sides (including the top) and hollow on the inside. How much wood, measured in cubic inches, did Sophia use to build the chest?

Problem 19

Jen, my dog, used to like eating my socks. From January to June she ate 54 of my socks. I know that each month she ate 2 more than the previous month. How many socks did she eat in June?

Problem 20

Samantha takes her younger cousins Billy and Tommy to buy a piece of candy at the store. The only money Billy brought is a few quarters and the only money Tommy brought is a few dimes. Billy tries to buy a piece of candy, but the clerk says, "Sorry Billy, you need 18 more cents to buy the candy." To show off, Tommy tries to buy the same piece of candy, and the clerk says, "You don't have enough money either Tommy, you need 8 more cents." Samantha bought the candy herself with a $1 bill and gave it to her cousins to share. How much did the piece of candy cost in cents?

1.5 ZIML February 2017 Division E

Below are the 20 Problems from the Division E ZIML Competition held in February 2017.
The answer key is available on p.162 in the Appendix.
Full solutions to these questions are available starting on p.110.

Problem 1
How many triangles are in the diagram below?

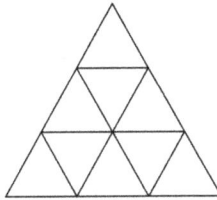

Problem 2
Corry went to the forest every day for a week to collect rocks. The first day he collected 5 rocks. Each of the following days he collected 2 more rocks than the previous day. What is the average of the number of rocks he collected each day?

Problem 3
There are 48 students in a class. If 3 more boys joined the class, the number of boys would be twice the number of girls. What is the current number of boys in the class?

Problem 4

The 'L' shape below is made up of four 1×1 squares.

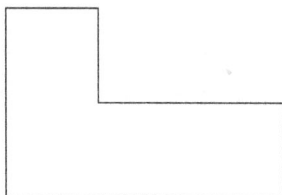

How many shapes of like the one above would you need to cover an 8×8 checkerboard? If it is not possible enter '0' as your answer.

Problem 5

The number 64 has the property that it is divisible by its units digit. How many whole numbers between 10 and 30 have this property?

Problem 6

Mary is baking a lot of cookies for a bake sale at Samantha's school. To bake the cookies she needs a total of 60 eggs. The eggs come in small cartons containing 6 eggs or large cartons containing 12 eggs. If Mary buys a total of 7 cartons, how many small cartons does she buy?

Problem 7

What is the smallest positive integer that is divisible by 2 and 3 that consists entirely of 2s and 3s, with at least one of each?

Problem 8

In the following diagram the length of AB is 6 and the length of AD is 3.

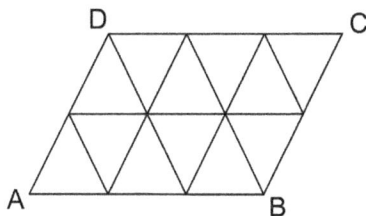

What is the perimeter of one of the small triangles?

Problem 9

Cody likes separating his M&M's by color before eating them. He has now only blue and green M&M's left. Before he ate 3 blue M&M's and 7 green M&M's, he had 50 M&M's. Now he has 2 more green M&M's than blue M&M's. How many green M&M's does he have left?

Problem 10

Teachers and students from the Areteem Summer Camp visited the museum. They bought a total of 99 tickets for 218 dollars. If each teacher ticket costs 4 dollars, and each student ticket costs 2 dollars, how many students were there?

Problem 11

ABCD is a parallelogram and E is the midpoint of DC. If the area of triangle AED is 4, what is the shaded area?

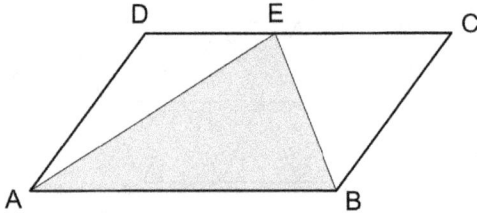

Problem 12

Three friends have a total of 6 identical pencils, and each one has at least one pencil. In how many ways can this happen?

Problem 13

Myles, Mya and Myron entered a team competition where they each had to run 400 meters and add up their times to get their team score. Myles was 4 seconds faster than Myron, and Mya was 10 seconds faster than Myron. If their total time was 181 seconds, how many seconds did it take Myron to complete the course?

Problem 14

A garden of rectangular shape is shown in the diagram. The shaded regions are grass, and the unshaded regions are empty spaces in the shape of four congruent regular hexagons.

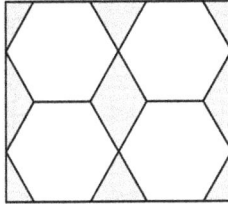

If the are of the grass portion is $25\,\text{ft}^2$, how many ft^2 of empty space are there?

Problem 15

The Little Twelve Basketball Conference has two divisions, with six teams in each division. Each team plays each of the other teams in its own division twice and every team in the other division once. How many conference games are scheduled?

Problem 16
The parallelogram $ABCD$ has area $48\,\text{cm}^2$, $AE = 8$ cm, $CE = 4$ cm.

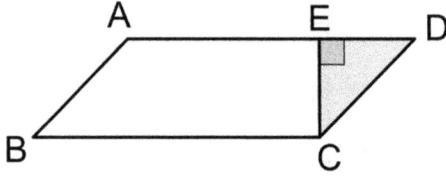

What is the area of the shaded region in cm^2?

Problem 17
Charles and David are fast typists. Charles can type 12 more words per minute than David. Charles started typing, and 2 minutes later David also started typing, and they both stopped after 3 more minutes. Given that they typed 780 words altogether, how many words did Charles type?

Problem 18
A certain fraction, $\dfrac{m}{n}$, if reduced to lowest terms, equals $\dfrac{5}{11}$. If $m+n = 80$, what is n?

Problem 19
A stop sign, which is a regular octagon, can be formed from at 1-foot square sheet of metal by making four straight cuts that snip off the corners. The resulting polygon will have side length $\sqrt{A} - B$ ft. What is B?

Problem 20

Debbie is mixing orange juice concentrate for her restaurant. The first juice concentrate is 64% real orange juice. The second is only 48% real orange juice. How many ounces of 48% real orange juice should she use to make 1600 ounces of 58% real orange juice?

1.6 ZIML March 2017 Division E

Below are the 20 Problems from the Division E ZIML Competition held in March 2017.
The answer key is available on p.163 in the Appendix.
Full solutions to these questions are available starting on p.118.

Problem 1
You are the first in line to buy tickets for a raffle at your school. They have tickets numbered 01, 02, 03, ..., 99. Your lucky number is 6, so you want to buy all the tickets that have exactly one 6. How many tickets will you buy?

Problem 2
Two equilateral triangles overlapped to form the star in the diagram below.

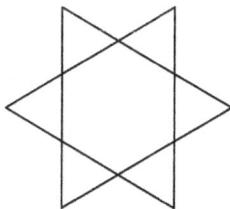

How many lines of symmetry does the star have?

Problem 3
What is the next number in the sequence $\frac{1}{11}$, 1, 11, 121, 1331, ...?

Problem 4
5 identical rectangles are combined into a larger rectangle as shown in the picture below.

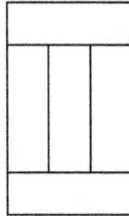

In each of the small rectangles, the small side is 2 inches long. How many inches is the perimeter of the big rectangle?

Problem 5
The ratio of girls to boys participating in intramural volleyball at Georgetown Middle School is 7 to 4. There are 42 girls in the program. What is the total number of participants?

Problem 6
What is the smallest counting number that is divisible by both 6 and 10, and is bigger than 100?

Problem 7
A pen and pencil together cost $2.10. The price of the pen is 6 times the price of the pencil. How many cents does the pencil cost?

Problem 8

You are writing a list of counting numbers on a piece of paper. You are counting in 3s starting with the number 103. If you wrote 7 numbers, what is the average of all of the numbers you wrote?

Problem 9

Lindy drew a triangle with base 3 inches and height 4 inches. Her friend Lucy wanted to draw a triangle that looked the same but had bigger area. If Lucy doubled the dimensions of the base and the height of Lindy's triangle, how many square inches are in the area of Lucy's triangle?

Problem 10

You have $20, and your friend has $25. If you want to have twice as much money as your friend, how many dollars should your friend give you?

Problem 11

The diagram shows a box that is open on the top.

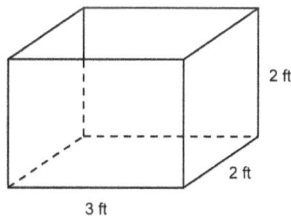

What is the surface area of the box in square feet?

Problem 12

A train leaves a station, traveling at 50 miles per hour. Two hours later, a second train leaves on a parallel track, traveling the same direction at 60 miles per hour. In how many hours will the second train catch up with the first train?

Problem 13

In the following diagram, four identical right triangles with legs of length 3 inches and 6 inches were put together to form a big square with a small square inside.

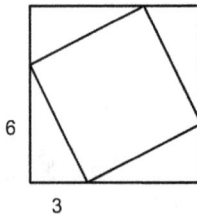

How many square inches are in the area of the small square?

Problem 14

Tom was bored while waiting for his mom to pick him up from school so he repeatedly wrote the number 1133557789 in a piece of paper producing the following pattern

$$11335577891133557789\ldots$$

When his mom finally came, he had written 144 digits in total. What was the last digit he wrote?

Problem 15
A 3×7 rectangle has perimeter 20. If a square is formed with the same perimeter, how much larger is the area of the square compared to the area of the rectangle?

Problem 16
Nine friends combined their money and bought a big bag of jellybeans. Each friend got the same amount of jellybeans. The label on the bag said how many jellybeans it contained, but the friends were only able to read the hundreds digit, which was 1, and the ones digit, which was 5 (only the tens digit was missing). How many jellybeans did each friend get?

Problem 17
Sixty vehicles (cars and motorcycles) are parked in a parking lot. In total there are 190 wheels. Given that a car has 4 wheels and a motorcycle has 2 wheels, how many cars are in the parking lot?

Problem 18

In the following figure the big square has side length 10 inches and the little squares have side length 5 inches.

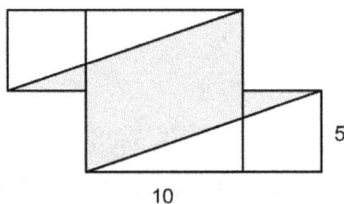

The shaded region has an area that is $N\%$ of the full figure, where N is a whole number. What is N?

Problem 19

Steve's bathtub can hold 96 gallons of water. The hot water tap takes 6 minutes to fill the tub, and the cold water tap takes 4 minutes to fill the tub. When the drain is not plugged, 8 gallons of water go down the drain every minute. If Steven opens both taps but forgets to put in the plug, how many minutes will it take to fill the tub?

Problem 20

A whole number smaller than 40 has the following properties: it leaves a remainder of 2 when you divide it by 5, it leaves a remainder of 1 when you divide it by 2, and leaves no remainder when you divide it by 3. What is the number?

1.7 ZIML April 2017 Division E

Below are the 20 Problems from the Division E ZIML Competition held in April 2017.
The answer key is available on p.164 in the Appendix.
Full solutions to these questions are available starting on p.125.

Problem 1
Suppose the full rectangle $ABCD$ shown below has area 40.

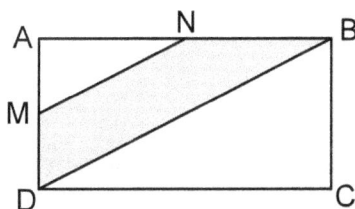

If M and N are in the exact middle of sides AD and AB, what is the area of the shaded region?

Problem 2
Calculate $54 + 67 + 33 + 84 + 46 + 64 + 16$.

Problem 3
At Ashland Middle School 40% of the students who play volleyball are boys. If 180 students (boys and girls) play volleyball, how many girls play volleyball?

Problem 4

Kyle and Lisa each bought some candies. If Kyle gave 4 candies to Lisa, they would both have the same number of candies. They pooled their candies together and counted, and there were 24 in total. How many candies did Kyle have originally?

Problem 5

Find the largest two-digit number that leaves a remainder of 2 when divided by 4 and 6.

Problem 6

Jerry lives in a city where all the roads are organized into square blocks and the roads either run North/South or East/West (forming a big square grid). Jerry goes out walking one day. He starts by walking 3 blocks North and then 4 blocks East. Jerry next walks 2 blocks South and 1 block West, when he realizes it is getting late and he must get home quickly. If Jerry walks on the roads home using as few blocks as possible, how many blocks will it take?

Problem 7

Adam, Bob, and Cynthia each draw the shapes shown in the diagram below (labeled A for Adam, B for Bob, and C for Cynthia).

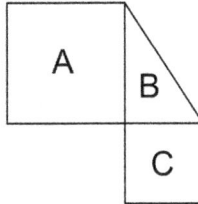

Adam and Cynthia drew squares, while Bob drew a triangle. If the area of Adam's square is 9 and the area of Bob's triangle is 3, what is the area of Cynthia's triangle? Round your answer to the nearest tenth if necessary.

Problem 8

If 8 people are in a room and everyone shakes everyone else's hand exactly once, how many handshakes were exchanged in the room?

Problem 9

Uncle Jim got lost while we were driving back to California from Montreal. He saw a sign that said how many miles away we were from Los Angeles.

Since he was driving fast, he couldn't quite see the number, but he knew it had 4 digits. I saw the number, but I wanted to have some fun so instead of telling him the number right away I gave him some clues:

(1) The number has the digit 1 somewhere.

(2) The digit in the hundreds place is three times the digit in the thousands place.

(3) The digit in the ones place is 4 times the digit in the tens place.

(4) The thousands digit is 2.

How far away are we from Los Angeles in miles?

Problem 10

In a math competition, there are 30 questions. Each correct answer earns 5 points. One point is taken away for each incorrectly answered or unanswered question. Jenny received 114 points. How many questions did she answer correctly?

Problem 11

How many edges are there in a cube?

Problem 12

Melisa drove for 3 hours at a rate of 50 miles per hour and for 2 hours at 60 miles per hour. What was her average speed for the whole journey in miles per hour?

Problem 13

Molly used 20 identical sticks to create the figure below, where all the vertical sticks are parallel and all the horizontal sticks are parallel.

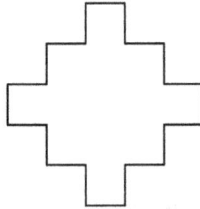

The total length of all the sticks Molly used is 100 cm. What is the area of the shape in square cm?

Problem 14

Suppose Chris can paint an entire house in twelve hours, and Bill can do it in eight hours. How long would it take the two painters working together to paint the house? Express your answer as a decimal in hours, rounded to the nearest tenth if necessary.

Problem 15

The product of the two 7-digit numbers 6060606 and 7070707 has hundreds digit A and tens digit B. What is the sum $A + B$?

Problem 16

Tim and Will have to type up a report that contains 600 words. Tim can type 50 words per minute, but Will can only type 40 words per minute. Will starts typing the report, and after some time Tim takes over to speed up the process. Together it takes a total of 13 minutes to type the report. How many minutes does Tim type?

Problem 17

Monkey George had a basket of bananas. The total weight including the bananas and the basket is 32 pounds. Monkey George ate half of the bananas, and then the total weight including the bananas and the basket was 17 pounds. How heavy is the basket in pounds?

Problem 18

Arrange several equilateral triangles, all of whose side lengths are 2 cm, to form a long parallelogram, as demonstrated in the diagram.

Assume the perimeter of the long parallelogram is 44 cm, how many triangles are there?

Problem 19

Paul wrote the word *SCIENCE* over and over in his notebook:

SCIENCESCIENCE ··· *SCIENCE*
SCIENCESCIENCE ··· *SCIENCE*
SCIENCESCIENCE ··· *SCIENCE*
⋮

Each line in his notebook has room for 50 letters, and Paul wrote as many complete *SCIENCE* words as he could in each line. If Paul wrote 80 copies of the word *SCIENCE* in total, how many lines did he use?

Problem 20

Susie and Frankie were building a tower together out of blocks. The blocks Susie was using had a volume of 2 cubic units. Frankie knows that the volume of 2 of her blocks is the same as the volume of 5 of Susie's blocks. When the tower is finished, it contains 20 of Susie's blocks and 10 of Frankie's blocks. What is the total volume of the tower in cubic units?

1.8   ZIML May 2017 Division E

Below are the 20 Problems from the Division E ZIML Competition held in May 2017.
The answer key is available on p.165 in the Appendix.
Full solutions to these questions are available starting on p.133.

Problem 1
The figure below is made up of equilateral triangles of different sizes.

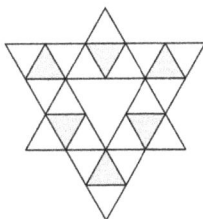

If the shaded triangles have combined area 12, what is the area of the whole figure?

Problem 2
How many multiples of 11 are there between 100 and 1000?

Problem 3
Quincey bought two boxes of pencils. One of the boxes was missing 4 pencils. If she has 62 pencils in total, how many pencils should each box have had?

Problem 4

Pat goes to a bakery to buy some cookies. The bakery sells chocolate chip, oatmeal, and peanut butter cookies. Pat wants to buy exactly 3 cookies. How many different assortments of the three cookies can Pat buy? Note: Pat does not buy the cookies in any particular order, so buying two oatmeal and one peanut butter is the same as one peanut butter and two oatmeal.

Problem 5

Some identical cubes were arranged in the corner of a room as in the diagram below.

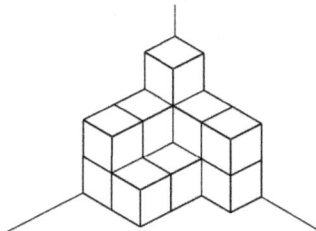

If there are no empty gaps between the wall and the cubes you can see, how many cubes were used in total?

Problem 6

Bob's dad was 25 years old when Bob was born. The age of Bob's dad last year was 3 times Bob's current age. What is the sum of the ages of Bob and his dad this year?

Problem 7

Four identical rectangles are arranged to form a big square like in the figure below.

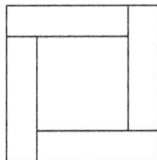

The area of the big square is 144, and the length of each of the rectangles is 10. What is the area of the small square inside the figure?

Problem 8

You and your brother were playing with some marbles. At the beginning you had 70 marbles, and your brother had 62. After playing for a while you lost some marbles (so they were given to your brother) and now your brother has 3 times as many marbles as you. How many marbles did you lose to your brother?

Problem 9

Sarah and her friends sit around a circular table, and start counting off numbers, clockwise, beginning with 1, and continue until 50 is counted. If there are between 10 and 20 people sitting around the table and the numbers 5 and 50 are counted by the same friend, how many friends are there in total?

Problem 10

In the diagram, each side is perpendicular to its adjacent sides, and all small sides have equal length.

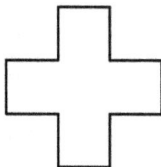

If the area of the whole figure is 180, what is the perimeter of the figure?

Problem 11

Rita bought some cat food. Her cat Melrose is picky when she eats, so Rita needs to buy different kinds of food. In total Rita bought 38 cans of food. She bought 4 more cans of Savory Shreds than Classic Pate, and 10 more cans of Prime Fillets than Classic Pate. If whe only bought cans of those three flavors, how many cans of Classic Pate did she boy?

Problem 12

You are setting up the passcode for your new phone. You have to choose 1 lowercase letter and 3 digits (you can repeat digits if you want). How many codes can you make if you only want to use odd digits and you do not want to use vowels? Recall there are 26 total letters, and 5 of them are vowels.

Problem 13

Kaileigh wants to create some figures using some toothpicks and small balls of clay. She will use the small balls of clay to glue together the toothpicks by placing the clay at the ends of the toothpicks. She wants to create a figure that looks as if she had two cubes glued together on one face. What is the sum of the number of toothpicks and the number of balls of clay she will need?

Problem 14

One drain pipe can empty a swimming pool in 6 hours. Another pipe takes 3 hours. If both pipes are used simultaneously to drain the pool, how long does it take to drain the pool? Give your answer in hours, rounded to the nearest tenth if necessary.

Problem 15

Two containers of 1-gallon each contain orange juice. One container is 70% full and the other is 80% full. You add water to fill each container completely, then pour the orange juice in both containers into one large container. What percent of the mixture in the large container is orange juice?

Problem 16

In the following diagram we have A squares and B rectangles.

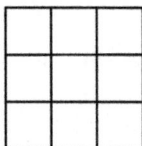

What is $B - A$? Remember that a square is also a rectangle.

Problem 17

Natalie wrote a list of 6 numbers on a piece of paper. She started the list with her two favorite numbers, the smaller one first. Each of the following numbers on Natalie's list was the sum of the two previous numbers on the list. If the last two numbers on the list are 504 and 816, what is the smaller of Natalie's favorite numbers?

Problem 18

Frank and Lacey were on a huge park with several ice cream shops inside it. They were trying to figure out where the closest one was. They found a sign that had the coordinates of some places of interest in the park, including the ice cream shops. The sign said they were currently standing on $(3,5)$, and listed ice cream shops on the points $(6,9)$, $(3,0)$, $(6,6)$ and $(4,0)$. What is the sum of the x-coordinate and y-coordinate of the closest ice cream shop?

Problem 19

Sam decided it was time to be healthy. He started doing crunches every morning and every day he increased the number of crunches he did. The first day he would do 10 crunches; the second day he would add 2 more, so he would do 12 crunches; and on the third day he'd add $2 \times 2 = 4$ more, so he'd do 16 crunches. Each of the following days he continued to add twice as many crunches as he added the previous day to his routine. If Sam followed this pattern for 5 days, how many crunches did he do on the fifth day?

Problem 20

Two friends leave the same place at the same time traveling in the same direction. One travels at a speed of 55 mph and the other travels at a rate of 65 mph. After 2 hours, how many miles will they be away from each other?

1.9 ZIML June 2017 Division E

Below are the 20 Problems from the Division E ZIML Competition held in June 2017.
The answer key is available on p.166 in the Appendix.
Full solutions to these questions are available starting on p.140.

Problem 1
Calculate
$$63 \times 25 \div 7 \times 2 \div 8 \div 5 \times 64 \div 9.$$

Problem 2
Julie buys lemons and limes at the grocery store. The price of lemons is 4 lemons per $1 and the price of limes is 6 limes for $1. Julie buys a total of 6 lemons and 21 limes. How much money does Julie spend in total? Express your answers in dollars, rounded to the nearest hundredth if necessary.

Problem 3

Kori is bored at her mom's office and started playing with some squared sticky notes she found on her desk. She made some figures in one of the walls that look like:

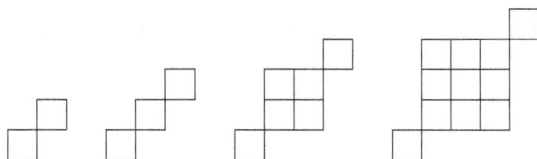

If she keeps making this kind of figures following the same pattern, how many sticky notes will she need for the 10th figure?

Problem 4

Divide a big square into 6 identical small rectangles, as shown.

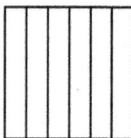

Suppose the big square has an area of 144 square inches. What is the perimeter of one of the 6 small rectangles?

Problem 5

If you add up all the positive factors of 36, what do you get? (Remember that both 1 and 36 are still factors of 36.)

Problem 6

You have two big buckets of water: one is red and the other one is blue. The red bucket contains 14 liters of water, and the blue bucket contains 18 liters of water. If you want the red bucket to contain 3 times as much water as the blue bucket, how many liters of water do you need to pour from the blue bucket into the red bucket?

Problem 7

Let A, B, C, D have coordinates $(-1, 1), (-1, 4), (2, 4), (2, 1)$. What is the area of quadrilateral $ABCD$?

Problem 8

Troy went to the store to buy some candy. He brought along a $10 bill, 3 quarters, 1 dime, 2 nickels, and 2 pennies. After picking out his candy, the total cost was $6.23. The cashier warned Troy that she only had $1 bills and pennies to give change. Troy does not like pennies, so he pays in such a way that he gets back the smallest number of pennies possible. How many pennies does Troy get back?

Problem 9

Teachers and students from the Areteem Summer Camp visited the museum. They bought a total of 99 tickets for 218 dollars. If each teacher ticket costs 4 dollars, and each student ticket costs 2 dollars, how many teachers were there?

Problem 10

Wally drew ten stacked 1×1 squares and then Molly drew a triangle as in the diagram below.

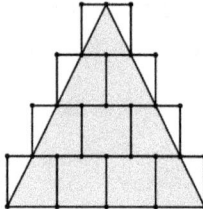

What is the area of Molly's triangle?

Problem 11

Cameron, Mandy and Taylor decided to put all their savings together to buy the new video game console that all the kids were playing. They had exactly enough money to buy the video game console that cost $300 and an extra controller that cost $20. If Taylor had $20 less than Mandy and Mandy had $50 more than Cameron, how much money (in dollars) did Mandy have? (Don't include the dollar sign ($) in your answer.)

Problem 12

Terry and Susan are entered in a 24-mile race. Susan's average rate is 4 miles per hour and Terry's average rate is 6 miles per hour. Both start at the same time. How many miles will Susan be away from the finish line when Terry crosses the line?

Problem 13

In the diagram below, there are 36 rectangular grid points, evenly spaced, and the distance between each pair of adjacent points is 1.

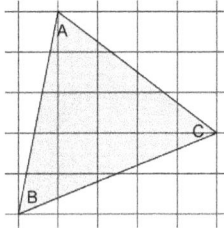

Find the area of $\triangle ABC$. Write your answer as a decimal rounded to the nearest tenth.

Problem 14

The table shows some of the results of a survey.

	Read	Don't Read	Total
Male	?	27	?
Female	99	?	165
Total	207	93	300

Using the above data, $K\%$ of the males surveyed read the newspaper. What is K? Round your answer to the nearest integer if necessary.

Problem 15
Given below is a list of the number of innings pitched by a player in the Major Baseball League:

$$5, 7, 4, 6, 8, 2, 4, 9$$

How many innings does this player have to pitch in the next game so that the average number of innings pitched in total is 6?

Problem 16
Suppose an uncle distributes $1 bills among his nieces and nephews. If he distributed the dollars evenly among the nieces, each niece would get 6 dollars, and if he distributed them evenly among the nephews, each nephew would get 10 dollars. If the uncle has less than 50 $1 bills in total, exactly how many $1 bills does he have?

Problem 17
Bob is building a model rectangular table with solid ends out of clay, as in the diagram below.

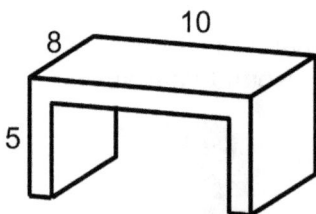

If the model is 10 inches by 8 inches by 5 inches and is 1 inch thick everywhere, what is the volume of clay (in cubic inches) Bob needs to build his model?

Problem 18

Stan can load the truck in 40 minutes. If I help him, it takes us 15 minutes. How many minutes will it take me to load the truck alone?

Problem 19

How many whole numbers between 99 and 999 have exactly one 0?

Problem 20

Arrange several squares, all of whose side lengths are 2 cm, to form a long rectangle, as shown in the diagram.

Assume the perimeter of the long rectangle is 144 cm, how many squares are there?

2. ZIML Solutions

This part of the book contains the official solutions to the problems from the nine Division E ZIML Contests from the 2016-17 School Year.

Students are encouraged to discuss and share their own methods to the problems using the Discussion Forum on ziml.areteem.org.

2.1 ZIML October 2016 Division E

Below are the solutions from the Division E ZIML Competition held in October 2016.

The problems from the contest are available on p.15.

Problem 1 Solution
By grouping we have $58 + 42 = 100$ and $33 + 17 = 50$ so the sum is $100 + 50 = 150$.

Answer: 150

Problem 2 Solution
The cows in total eat $3 \times 16 = 48$ bundles of hay per day. Similarly the horses eat $4 \times 14 = 56$ bundles of hay per day. Hence in total they eat $48 + 56 = 104$ bundles of hay in a single day. Since there are 7 days in a week, the farmer needs $7 \times 104 = 728$ bundles of hay each week.

Answer: 728

Problem 3 Solution
First note $0.48 = \dfrac{48}{100}$. We have the greatest common factor of 48 and 100 is 4, so dividing the numerator and denominator by 4 we have $\dfrac{48}{100} = \dfrac{12}{25}$. Thus, the denominator $M = 25$.

Answer: 25

Problem 4 Solution
Connect A to C and B to D. Note this divides the rectangle into 8 triangles that are all the same. Since half of these triangles are shaded, the shaded region has area $30 \div 2 = 15$.

Answer: 15

Problem 5 Solution
Work one digit at a time. The ones digit is $2 \times 2 = 4$ and then the hundreds digit is $2 \times 4 = 8$.

Answer: 824

Problem 6 Solution
Each line can intersect the circle in two places, and the lines can intersect each other once. This gives a total of $2 + 2 + 1 = 5$ points of intersection.

Answer: 5

Problem 7 Solution
Note that $100 \div 8$ is 12 with remainder 4. (Alternatively note $12 \times 8 = 96$ is the largest multiple of 8 less than 100.) Hence Josh buys 12 pairs of socks (for 96 dollars) and gets 4 dollars back in change.

Answer: 4

Problem 8 Solution
The number decreases by 4 each time. Note the 3rd number is $-4 = -4 \times 1$, the 4th number is $-8 = -4 \times 2$, etc., so the 20th number is $-4 \times 18 = -72$.

Answer: -72

Problem 9 Solution
The basket and sandwich weight decreased by $19 - 12 = 7$ lbs when half of the sandwiches were eaten. Thus in total the sandwiches weight $2 \times 7 = 14$ lbs. The remaining $19 - 14 = 5$ lbs must be the weight of the basket.

Answer: 5

Problem 10 Solution

Think of them shaking hands one person at a time. The first person must share hands with the other 7 people. The next person only needs to shake hands with 6 people, since they have already shaken the first person's hand. This pattern continues, so there are $7 + 6 + 5 + 4 + 3 + 2 + 1 = 28$ handshakes in total.

Answer: 28

Problem 11 Solution

The shaded region is the two squares minus the unshaded triangle with base 6 and height 4, so the area is $4^2 + 2^2 - \dfrac{1}{2} \times 4 \times 6 = 8$.

Answer: 8

Problem 12 Solution

Since the trains travel in opposite directions, after one hour they will be $56 + 64 = 120$ miles apart. Hence in three hours they will be $3 \times 120 = 360$ miles apart.

Answer: 360

Problem 13 Solution

Note $3 - 4 = -1$ then $3 - 3 = 0$ then $4 - 3 = 1$. Continuing $6 - 4 = 2, 9 - 6 = 3, 13 - 9 = 4, 18 - 13 = 5$. Thus the numbers differ by $-1, 0, 1, 2, 3, 4, 5$. Hence the next number will differ from 18 by 6, so must be $18 + 6 = 24$.

Answer: 24

Problem 14 Solution

Removing 10 basketballs we have 40 total balls and an equal number of basketballs and volleyballs. Thus there are $40 \div 2 = 20$ volleyballs.

Answer: 20

Problem 15 Solution

There are 12 edges, 8 corners, and 6 faces in a cube, so the sum is $12 + 8 + 6 = 26$.

Answer: 26

Problem 16 Solution

123456789 consists of 9 numbers. Therefore the 9th, 18th, 27th, etc. number in the pattern will all be 9. Since $1009 \div 9 = 112$ remainder 1, the 1008th number is 9 and the 1009th number is 1.

Answer: 1

Problem 17 Solution

We know that $900 = 30 \times 30$. Trying $30 \times 31 = 930$, $31 \times 32 = 992$, and then $32 \times 33 = 1056$ we see 32 is the smaller of the two page numbers.

Answer: 32

Problem 18 Solution

Beth has 3 times as much as Cynthia and so Alice has $3 \times 2 = 6$ times as much as Cynthia. Therefore, the 50 dollars is $1 + 3 + 6 = 10$ times more money than Cynthia has. Cynthia thus has $50 \div 10 = 5$ dollars and hence Beth has 15 dollars.

Answer: 15

Problem 19 Solution

The area of each smaller circle is $1/4$ of the large circle, so the sum of the areas of the smaller circles equals the area of the large circle. Thus the overlap regions of the smaller circle have the same area as the regions between the large circle and the small circles. Consequently, considering the partially shaded small circle, the missing portion has the same area as the region outside this small circle. Therefore the area of the shaded region equals the area of a small circle, which is $1/4$ of the large circle, hence the answer is 25.

Answer: 25

Problem 20 Solution

Note $90 \div 5 = 18$. If a student gets only 17 questions correct they can get at most $17 \times 5 + 3 \times 1 = 88$ points (if they get 17 correct and leave 3 unanswered). Hence a student must get at least 18 questions correct. If a student gets 18 questions correct, they can get either 90, 91, or 92 points depending on whether the remaining two questions are incorrect or unanswered. If a student gets 19 questions correct, they can get either 95 or 96 points. Lastly a student gets 100 points if they answer all 20 questions correctly. Thus there are 6 possible scores ≥ 90.

Answer: 6

2.2 ZIML November 2016 Division E

Below are the solutions from the Division E ZIML Competition held in November 2016.
The problems from the contest are available on p.21.

Problem 1 Solution
Since Lucas was highlighting 450 words, it took him $450 \times 2 = 900$ seconds to highlight those words. Since there are 60 seconds in one minute, it took him $900 \div 60 = 15$ minutes in total.

Answer: 15

Problem 2 Solution
Scott's entire candy bar has area

$$3 \times 6 = 18.$$

Thus half the candy bar is $18 \div 2 = 9$ square inches. Mike's candy bar has area

$$3 \times 8 = 24,$$

so he must eat

$$\frac{9}{24} = \frac{3}{8}$$

of his candy bar. We have

$$\frac{3}{8} = 0.375 = 37.5\%$$

so our answer is 37.5.

Answer: 37.5

Problem 3 Solution
Note the pattern is $+2, -1, +2, -1, \ldots$. Therefore every other number increases by 1. The 2nd, 4th, 6th, 8th, etc numbers in the

pattern are
$$7, 8, 9, 10, \ldots$$
so the 20th number in the original pattern is $7 + 10 - 1 = 16$. For reference, the first 20 numbers in the pattern are

$$5, 7, 6, 8, 7, 9, 8, 10, 9, 11, 10, 12, 11, 13, 12, 14, 13, 15, 14, 16$$

Answer: 16

Problem 4 Solution
Note we can cancel the 2's, the 3's, up to the 9's. Hence the above simplifies to $\frac{1}{10}$ with numerator $N = 1$ and denominator $M = 10$. Thus $N + M = 1 + 10 = 11$.

Answer: 11

Problem 5 Solution
Note on the front of the wall Little Pig needed to paint

$$4 \times 5 = 20$$

faces. Similarly he painted 20 faces on the back of the wall. The top of the wall has 5 faces, while the left and right of the wall each have 4 faces. This gives a total of

$$2 \times 20 + 5 + 2 \times 4 = 40 + 5 + 8 = 53$$

faces that Little Pig had to paint.

Answer: 53

Problem 6 Solution
Grouping we have $(-5 + 10) + (-15 + 20) + \cdots + (-95 + 100) = 10 \times 5 = 50$.

Answer: 50

Problem 7 Solution

There is one line of symmetry through each vertex of the pentagon. The 5 lines of symmetry are shown below:

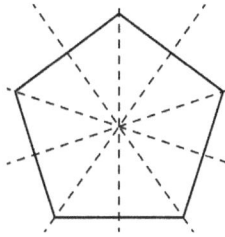

Answer: 5

Problem 8 Solution

It is most efficient here to use as many $7 bills as possible. We have $41 \div 7 = 5$ with remainder 6. Since $6 = 4 + 2$ we can use five $7 bills, one $4 bill, and one $2 bill for 7 bills in total.

Answer: 7

Problem 9 Solution

If David gave 4 cherries to Ed, they would each have $82 \div 2 = 41$ cherries. Thus David picked $41 + 4 = 45$ cherries.

Answer: 45

Problem 10 Solution

To be a four-digit number, it must start with either 1 or 3. If it starts with a 1, the 3 must appear in either the hundreds, tens, or ones place, so there are 3 possibilities ($1300, 1030, 1003$. Similarly there are 3 possibilities starting with a 3 ($3100, 3010, 3001$) for $3 + 3 = 6$ possibilities in total.

Answer: 6

Problem 11 Solution

We want the length l and width w to be as close together as possible. We know that

$$2l + 2w = 18$$

so $l + w = 9$. Hence if the length and width are 5 and 4, we get the largest possible area:

$$5 \times 4 = 20.$$

Answer: 20

Problem 12 Solution

Note the pattern repeats every 8 terms. The 8 terms are:

$$\leftarrow, \; \rightarrow, \; \uparrow, \; \downarrow, \; \rightarrow, \; \leftarrow, \; \downarrow, \; \uparrow.$$

Since $18 \div 8 = 2$ with remainder 2, the sequence $\leftarrow, \; \rightarrow$ occurs 3 times in the first 18 terms.

Answer: 3

Problem 13 Solution

We know that Nathan is $23 - 17 = 6$ years older than Matt. Since the sum of their ages now is 50, we know that twice of Matt's age is $50 - 6 = 44$. Therefore Matt is $44 \div 2 = 22$ years old.

Answer: 22

Problem 14 Solution

A perfect score is $10 \times 5 = 50$ points. Olivia's score is $50 - 29 = 21$ less than a perfect. Since each incorrect answer reduces her score by $5 + 2 = 7$ points, she must have incorrectly answered

$21 \div 7 = 3$ questions. Hence Olivia answered $10 - 3 = 7$ questions correctly.

Answer: 7

Problem 15 Solution
William's pool has a volume of

$$10 \times 15 \times 6 = 900$$

cubic feet. Therefore William's pool holds

$$900 \times \frac{75}{10} = 90 \times 75 = 6750$$

gallons of water.

Answer: 6750

Problem 16 Solution
The next palindrome is 3223 so the product of the digits is $3 \times 2 \times 2 \times 3 = 36$.

Answer: 36

Problem 17 Solution
First note we have $5 \times 5 = 25$ unit squares. Then there are $4 \times 4 = 16$ squares with area 4. Following this pattern, there are $3 \times 3 = 9$ squares with area 9, $2 \times 2 = 4$ squares with area 16, and $1 \times 1 = 1$ square with area 25. This gives a total of

$$25 + 16 + 9 + 4 + 1 = 55$$

squares.

Answer: 55

Problem 18 Solution
1 appears 10 times as the ones digit (in the numbers 1, 11, 21, 31, 41, ..., 91), 10 times as the tens digit (in the numbers 10, 11, 12,

13, ..., 19), and 1 time as the hundreds digit (in 100). Thus it appears $10 + 10 + 1 = 21$ times.

Answer: 21

Problem 19 Solution
Note if he pulls 12 socks without looking it is possible to get 4 red, 4 white, and 4 blue socks (so there are not 5 of the same color). However, once he picks the 13th sock he is guaranteed to have at least 5 of one color. Hence the answer is 13.

Answer: 13

Problem 20 Solution
Note that in the diagram below we have divided the big triangle into 9 of the small triangles.

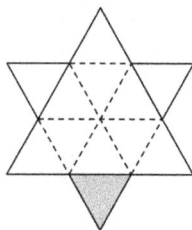

Hence the big triangle has area $4 \times 9 = 36$.

Answer: 36

2.3 ZIML December 2016 Division E

Below are the solutions from the Division E ZIML Competition held in December 2016.
The problems from the contest are available on p.27.

Problem 1 Solution
Note that

$$25 \times 3 \times 7 \times 4 = (25 \times 4) \times (3 \times 7) = 100 \times 21 = 2100.$$

Answer: 2100

Problem 2 Solution
The ones digit is the second largest one digit number, so it is 8. From here we can calculate the remaining digits one by one. So, the tens digit is $8 \div 2 = 4$, the hundreds digit is $4 \div 2 + 1 = 3$, and the thousands digit is $4 - 3 = 1$. Thus, the number is 1348.

Answer: 1348

Problem 3 Solution
Consider the diagram below, where the full square is divided into 8 identical triangles.

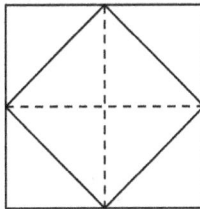

As the inner square is made up of 4 of these triangles, each triangle has area $20 \div 4 = 5$ cm^2.

Answer: 5

Problem 4 Solution

First we convert minutes to hours. Since there are 60 minutes in 1 hour, we have

$$90 \text{ minutes} = \frac{90}{60} = \frac{3}{2} \text{ hours.}$$

Similarly

$$30 \text{ minutes} = \frac{30}{60} = \frac{1}{2} \text{ hours.}$$

Therefore, in total Melissa travels

$$\frac{3}{2} \times 50 + \frac{1}{2} \times 60 = 75 + 30 = 105$$

miles.

Answer: 105

Problem 5 Solution

Note using the diagram we see that the shorter side of the smaller rectangles must be half the length of the longer side. Hence the shorter side of each is $4 \div 2 = 2$ inches. Therefore the sides of the large rectangle are 4 and $4 + 2 = 6$ inches so it has an area of $4 \times 6 = 24$ square inches.

Answer: 24

Problem 6 Solution

If Johnny found $48 dollars worth of gold when it was worth $16 an ounce, Johnny found

$$48 \div 16 = 3$$

ounces of gold. This gold is now worth $3 \times 300 = 900$ dollars.

Answer: 900

Problem 7 Solution
Grace has blocks that are 10 cm high, so she can build towers that have a height that is a multiple of 10. Similarly, Andrew can build towers that have a height that is a multiple of 6. Multiples of 10 are

$$10, 20, 30, 40, \ldots$$

while multiples of 6 are

$$6, 12, 18, 24, 30, 36, \ldots$$

Therefore, 30 cm is the smallest height that Grace and Andrew can both build.

Answer: 30

Problem 8 Solution
Amy can mow

$$120 \div 30 = 4$$

square feet of the lawn per minute. When Amy and Claire work together, they can finish mowing the lawn in one hour. This means they can mow

$$420 \div 60 = 7$$

square feet per minute. Therefore, Claire can mow

$$7 - 4 = 3$$

square feet per minute.

Answer: 3

Problem 9 Solution

Since the first and fourth trees are 60 feet apart, each tree is $60 \div 3 = 20$ feet away from the next tree. Hence the first and last tree are $5 \times 20 = 100$ feet apart.

Answer: 100

Problem 10 Solution

In total there are

$$2+3+4+5+6+7 = 27$$

stars in the equation. Therefore

$$27 \times \bigstar = 2700,$$

so we see that $\bigstar$ must be 100.

Answer: 100

Problem 11 Solution

If all of them were motorcycles, there would be a total of

$$7 \times 2 = 14$$

wheels. This is $18 - 14 = 4$ less than the actual number of wheels. As each car has 2 extra wheels, there must be 2 cars.

Answer: 2

Problem 12 Solution

If the circles all intersect, we can divide them into 7 regions as in the diagram below

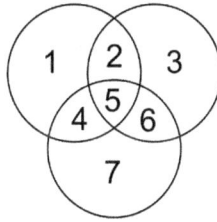

Answer: 7

Problem 13 Solution

Continuing the pattern the branches have

$$1, 1, 2, 3, 5, 8, 13, 21, 34, 55$$

leaves, so the tenth branch has 55 leaves.

Answer: 55

Problem 14 Solution

In inches, the box is $2 \times 12 = 24$ inches long, 12 inches wide, and 6 inches high. Therefore,

$$24 \div 2 = 12$$

cubes fit inside the box lengthwise,

$$12 \div 2 = 6$$

fit widthwise, and

$$6 \div 2 = 3$$

fit upwards. Hence a total of

$$12 \times 6 \times 3 = 72 \times 3 = 216$$

cubes fit in the box.

Answer: 216

Problem 15 Solution

If we pair off plants that the squirrels ate with the surviving plants, after we pair of all the plants eaten by squirrels we still have 61. This means that we must have paired off

$$121 - 61 = 60$$

plants so $60 \div 2 = 30$ of them were eaten by the squirrels.

Answer: 30

Problem 16 Solution

Bill's triangle has a base of 5. The base of Will's triangle is 10, which is double the length of Bill's triangle. Therefore the height of Will's triangle is also double the height of Bill's triangle. Thus the missing height is $2 \times 4 = 8$.

Answer: 8

Problem 17 Solution

Looking at the 2-digit multiples of 11, they are

$$11, 22, 33, 44, 55, 66, 77, 88, 99.$$

All of these numbers leave remainder of 0 when divided by 11. Adding 4 to each of these, the numbers

$$15, 26, 37, 48, 59, 70, 81, 92, 103$$

all leave remainder of 4 when divided by 11. Note that while 99 is less than 100, 103 is not. Hence the largest 2-digit number that leaves a remainder of 4 when divided by 11 is 92.

Answer: 92

Problem 18 Solution

Rotated we get

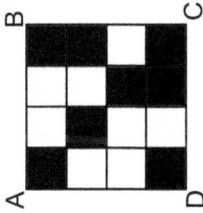

We see that 5 squares in the top half are now shaded black.

Answer: 5

Problem 19 Solution

Let's rewind time and pretend no ornaments are broken. Also, let's put aside for now 4 of the red ornaments so we would have $200 - 4 = 196$ ornaments in total. This way the number of red ornaments is exactly the same as 3 times the number of white ornaments. It seems that there are 14 more white ornaments than blue ornaments. Let's pretend we have 14 more blue ornaments, so we have $196 + 14 = 210$ ornaments in total and the same amount of blue and white. This way, we can make groups of $3 + 1 + 1 = 5$ ornaments (3 red, 1 white and 1 blue). We can make $210 \div 5 = 42$ groups of 5 ornaments, so we have 42 white, 42 blue and $42 \times 3 = 126$ red. Since we assumed we had 14 more blue ornaments and we set aside 4 red ornaments, we actually have 42 white, $42 - 14 = 28$ blue and $126 + 4 = 130$ red ornaments.

Answer: 130

Problem 20 Solution

We know either Brandon or Cathy drives. If Brandon drives, then Alex must sit in the front passenger seat and Cathy and David sit in the back seats. Either Cathy sits on the left and David on the right, or David on the left and Cathy on the right. Hence there are

2 seating arrangements if Brandon drives:

$$\begin{matrix} B & A \\ C & D \end{matrix} \quad \text{or} \quad \begin{matrix} B & A \\ D & C \end{matrix}$$

If Cathy drives, then Brandon and Alex must sit in the back seats and David in the front passenger seat. This again gives 2 more seating arrangements:

$$\begin{matrix} C & D \\ A & B \end{matrix} \quad \text{or} \quad \begin{matrix} C & D \\ B & A \end{matrix}$$

Hence there are $2 + 2 = 4$ total seating arrangements.

Answer: 4

2.4 ZIML January 2017 Division E

Below are the solutions from the Division E ZIML Competition held in January 2017.
The problems from the contest are available on p.35.

Problem 1 Solution
Since all the numbers are close to 90, we can use the following trick:

$$88 + 86 + 91 + 92 + 87 + 90 + 89 + 93 + 92 + 88$$
$$= 90 \times 10 - 2 - 4 + 1 + 2 - 3 + 0 - 1 + 3 + 2 - 2$$
$$= 896$$

This allows us to quickly calculate the sum.

Answer: 896

Problem 2 Solution
Let us make up our own 'unit' and suppose the height of a hobbit is one unit. Then an elf is three units tall, or equivalently, an elf is 2 units taller than a hobbit. Since we know the elf is 60 inches taller, our unit must be

$$60 \div 2 = 30 \text{ inches.}$$

Thus the elf is
$$3 \times 30 = 90$$
inches tall.

Answer: 90

Problem 3 Solution
There is one vertical, one horizontal, and two diagonal lines of symmetry as shown in below.

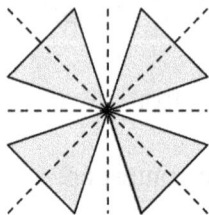

Answer: 4

Problem 4 Solution
Timmy has 3 choices for jeans. For each of these choices, he
has 8 choices for a shirt. Then he also has 2 different choices for
shoes. Since he can combine jeans, shirts, and shoes in any way,
there are a total of

$$3 \times 8 \times 2 = 48$$

different ways he can get dressed.

Answer: 48

Problem 5 Solution
We know that all the digits are among $1, 2, 3, \cdots, 9$. For the
difference between the first and last digit to be 8, one of the digits
must be 1 and the other must be 9. However, if the first digit
is 9, it is impossible for the 2nd digit to be twice the first digit.
Therefore we see the first digit is 1, the second is 2, the third is 4,
and the last is 9. Hence the number is 1249.

Answer: 1249

Problem 6 Solution
Since David gives out 25% of the fruit snacks each day, 75% of
the fruit snacks remain at the end of each day. So after two days,

$$75\% \times 75\% = \frac{3}{4} \times \frac{3}{4} = \frac{9}{16}$$

of the fruit snacks remain. We know this is equal to 36 fruit snacks. To find the original amount, we calculate

$$36 \div \frac{9}{16} = 64,$$

so there were originally 64 fruit snacks in the bag.

Answer: 64

Problem 7 Solution
Note the two shaded regions combine to form half a 2 by 2 square. Thus the area is 2.

Answer: 2

Problem 8 Solution
Note that the sum

$$80 + 70 + 50 = 200$$

counts twice the amount of diamonds found by Captain Hook, Popeye, and Sinbad. Therefore in total the three found

$$200 \div 2 = 100$$

diamonds. Since Popeye and Sinbad found 70 diamonds together, the remaining

$$100 - 70 = 30$$

were found by Captain Hook.

Answer: 30

Problem 9 Solution
Consider how many stickers are used in the ones, tens, and hundreds digits separately. For the ones digit, we use stickers in

$$2, 12, 22, 32, \ldots, 142,$$

a total of 15 times. In the tens digit, we use stickers in

$$20, 21, 22, \ldots, 29$$

and also

$$120, 121, 122, \ldots, 129$$

for an additional 20 times. Since there are no stickers needed for the hundreds digit, we use a total of

$$15 + 20 = 35$$

stickers.

Answer: 35

Problem 10 Solution

We are given that the rates that the two pipes fill the pool are in ratio $1.5 : 1 = 3 : 2$. Hence, the amount of water of the pool they fill in 10 hours is in the same ratio $3 : 2$. Therefore, in 10 hours, the first pipe fills $\frac{3}{5}$ of the pool, while the second pipe fills $\frac{2}{5}$ of the pool. Therefore, the slower of the two pipes can fill

$$\frac{2}{5} \div 10 = \frac{1}{25}$$

of the pool in a single hour. Hence it takes this pipe 25 hours to fill the pool alone.

Answer: 25

Problem 11 Solution

The 12×12 square can be divided into nine 4×4 squares as shown in the diagram below.

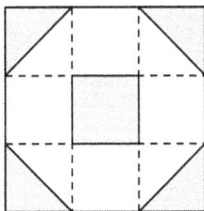

We see that the four shaded triangles are each half of one of these squares, so in total 3 squares worth of area is shaded. Hence the area of the shaded region is

$$3 \times (4 \times 4) = 48.$$

Thus the unshaded region has area

$$12 \times 12 - 48 = 144 - 48 = 96.$$

Answer: 96

Problem 12 Solution

First, we need to find the total distance the truck travels, which is

$$30 + 90 + 50 = 170 \text{ miles.}$$

Then the total hours the truck traveled is

$$1 + 2 + 1 = 4 \text{ hours.}$$

So, the average speed of the truck is

$$170 \div 4 = 42.5 \text{ miles per hour.}$$

Answer: 42.5

Problem 13 Solution

Number the months (January, February, to December) 1, 2, to 12. William's uncle comes every 2 months, so he will come on months

$$2, 4, 6, 8, 10, 12.$$

Similarly his grandmother will come on months

$$3, 6, 9, 12.$$

Combining these lists we see that neither visit on months

$$1, 5, 9, 11,$$

so there are 4 months where William is not visited.

Answer: 4

Problem 14 Solution

Note that every time the 5×3 rectangle is rolled, it moves either 5 units or 3 units, alternating between the two. Hence after 2 rolls, the rectangle moves $3 + 5 = 8$ units. Repeating we see the rectangle moves a total of

$$5 \times 8 = 40$$

units in 10 rolls.

Answer: 40

Problem 15 Solution

Note that the diagram has the same perimeter as a square with the same dimensions. The square thus has a side length of

$$108 \div 4 = 27$$

cm. As a side of the square is made up of 3 small sides from the diagram, each small side has length

$$27 \div 3 = 9$$

cm. As the diagram is made up of five squares with this side length, the total area of the shape is

$$5 \times 9^2 = 405\text{cm}^2.$$

Answer: 405

Problem 16 Solution
We know there are 11 animals in total. We first know that Thomas shows Thelma 5 tigers, so Thelma sees at least 5 tigers. For her to see twice as many tigers as lions, she thus must see at least 6 tigers and 3 lions. Note if the last two animals are hidden, then all the above statements are correct. If we want the maximum number of lions, suppose the hidden animals are both lions, for a total of 5 lions.

Answer: 5

Problem 17 Solution
20 kg of a snack that is $4.00 per kg should cost

$$20 \times 4.00 = 80$$

dollars in total. If the store owner uses 20 kilograms from only raisins, the 20 kg would only cost,

$$20 \times 3.50 = 70$$

dollars. This is

$$80 - 70 = 10$$

cheaper than it should be. Each kilogram of peanuts costs

$$4.75 - 3.50 = 1.25$$

more than a kilogram of raisins. Thus, there must be

$$10 \div 1.25 = 8$$

kilograms of peanuts. The remaining

$$20 - 8 = 12$$

kilograms are raisins.

Answer: 12

Problem 18 Solution
The full chest has volume

$$8 \times 5 \times 4 = 160 \text{ cubic inches.}$$

The hollow inside of the chest is $8 - 2 = 6$ inches long, $5 - 2 = 3$ inches wide, and $4 - 2 = 2$ inches tall, so has volume

$$6 \times 3 \times 2 = 36 \text{ cubic inches.}$$

Hence Sophia needed

$$160 - 36 = 124$$

cubic inches of wood to build the box.

Answer: 124

Problem 19 Solution
Since Jen eats 2 more socks each month, we know that the total of the socks she ate in January plus June is the same as February plus May, and is also the same as March plus April. Therefore Jen must have ate a total of

$$54 \div 3 = 18$$

socks in March and April combined. Since we also know Jen ate 2 more socks in April, she must have ate 8 socks in March and 10 in April. Working from here we see Jen ate

$$4, 6, 8, 10, 12, 14$$

socks in each of the months. This means Jen at 14 socks in June.

Answer: 14

Problem 20 Solution

Samantha paid with a $1 bill, so we know the candy is less than $1 or 100 cents. Since Billy has only quarters, he has either

$$25, 50, \text{ or } 75$$

cents. If he is 18 cents short of buying the candy, the price of the candy (in cents) must be one of

$$25 + 18 = 43, 50 + 18 = 68, \text{ or } 75 + 18 = 93$$

cents. Tommy has only dimes so the amount of money he has is

$$10, 20, 30, 40, 50, 60, 70, 80, \text{ or } 90$$

cents. He is 8 cents short of buying the candy, so the candy must be

$$18, 28, 38, 48, 58, 68, 78, 88, \text{ or } 98$$

cents. Hence, the candy must be 68 cents, as this is the only price that fits with both possibilities.

Answer: 68

2.5   ZIML February 2017 Division E

Below are the solutions from the Division E ZIML Competition held in February 2017.
The problems from the contest are available on p.43.

Problem 1 Solution
If we assume that the side of each of the small triangles is 1, we have 9 triangles of side length 1, 3 triangles of side length 2 and 1 triangle ofside length 3. So we have $9 + 3 + 1 = 13$ triangles in total.

Answer: 13

Problem 2 Solution
We can list the number of rocks he collected each day of the week: $5, 7, 9, 11, 13, 15$ and 17. Since this numbers form an arithmetic sequence and we have an odd number of terms, the average of them will be the number in the middle: 11.

Answer: 11

Problem 3 Solution
Let's pretend we have 3 more boys in the class, so we have $48 + 3 = 51$ students in total, and we have twice as much boys as girls. This means that for every 3 students in the class, 2 of them will be boys. So, we have $51 \div 3 \times 2 = 34$ boys. Since we pretended we had 3 more boys, the current number of boys is $34 - 3 = 31$.

Answer: 31

Problem 4 Solution
Note that we can fit 2 'L' shapes to form a 2×4 rectangle, and we can cover the 8×8 square with 8 of this rectangles. So, we will need $8 \times 2 = 16$ 'L' shapes.

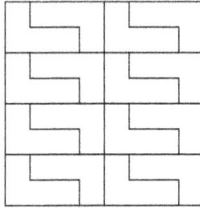

Answer: 16

Problem 5 Solution
It will be easier to list them based on their last digits: 11, 21; 12, 22; 24; 15, 25. So we have 7 whole numbers between 10 and 30 that are divisible by their units digit.

Answer: 7

Problem 6 Solution
If Mary bought 7 small cartons of eggs, she would have a total of

$$7 \times 6 = 42$$

eggs, which is

$$60 - 42 = 18$$

less than she needs. Hence some of the cartons must be large cartons. Since a large carton contains

$$12 - 6 = 6$$

extra eggs, if Mary replaces

$$18 \div 6 = 3$$

of the small cartons with large cartons she will have the correct number of eggs. Therefore Mary buys

$$7 - 3 = 4$$

small cartons of eggs and 3 large cartons of eggs.

Answer: 4

Problem 7 Solution

It is divisible by 2, so the last digit must be 2. It is divisible by 3, so the sum of the digits is a multiple of 3, thus there should be at least three 2s. The digit 3 has to appear once, and the smallest possible position for this 3 is the tens place. Therefore, the number is 2232.

Answer: 2232

Problem 8 Solution

Note that each of the triangles in the diagram is an isosceles triangle with one side of length $6 \div 3 = 2$ and two sides of length $3 \div 2 = 1.5$. So, the perimeter of each of the triangles is $2 + 1.5 + 1.5 = 5$.

Answer: 5

Problem 9 Solution

After eating 3 blue M&M's and 7 green M&M's, Cody has $50 - 3 - 7 = 40$ M&M's left. If he had 2 more blue M&M's, he would have $40 + 2 = 42$ M&M's in total and he would have the same amount of each color. That means he actually has $42 \div 2 = 21$ green M&M's.

Answer: 21

Problem 10 Solution

If instead we have 99 unsupervised students with no teachers, the 99 tickets would cost a total of

$$99 \times 2 = 198$$

dollars. This is

$$218 - 198 = 20$$

dollars cheaper than the actual 99 tickets. If we remove one student ticket and add one teacher ticket, it costs $2 extra. Since $20 extra was paid over the price of 99 student tickets, there must be

$$20 \div 2 = 10$$

teachers. Hence there are

$$99 - 10 = 89$$

students.

Answer: 89

Problem 11 Solution
Note that we can split the parallelogram $ABCD$ into four triangles with the same area by drawing a parallel line to AD through E. The shaded area is comprised of exactly two of those triangles, each of area 4, so it has area $2 \times 4 = 8$.

Answer: 8

Problem 12 Solution
As we know that each of them has at least one pencil, we just need to figure out in how many ways we can give the remaining 3 pencils to them. We will consider different cases: (1) Giving each of them 1 pencil can be done in one way. (2) Giving all three pencils to one single person can be done in 3 ways (choose which person gets the pencils). (3) Each having a different amount of pencils (so one has 0, one has 1 and one has 2) can be done in $3 \times 2 \times 1 = 6$ ways. Thus, we can give them the pencils in $1 + 3 + 6 = 10$ different ways.

Answer: 10

Problem 13 Solution

If Myles had taken 4 more seconds to complete the course, and Mya had taken 10 more seconds to complete the course, they would have all spent the same time and would have had a total time of $181 + 4 + 10 = 195$ seconds. That way, each of them would have taken $195 \div 3 = 65$ seconds to complete the course. Hence, it took Myron 65 seconds to complete the course.

Answer: 65

Problem 14 Solution

A hexagon can be broken down into 6 equilateral triangles. From here we can see that the unshaded region has the area of $4 \cdot 6 = 24$ of these equilateral triangles. The shaded region has the area of $1 \cdot 2 + 4 \cdot 1 + 4 \cdot \frac{1}{2} = 8$ equilateral triangles. So, for each shaded triangle we have 3 unshaded triangles. This means the unshaded area is 3 times the shaded area, that is, there are $3 \times 25 = 75 \, \text{ft}^2$ of empty space.

Answer: 75

Problem 15 Solution

Each of the 6 teams in each division will play 10 games against someone in their division (2 games against each of the 5 teams), so there will be $12 \times 10 \div 2 = 60$ such games. And each of the teams will have play 6 games against someone in the other division, so there will be $12 \times 6 \div 2 = 36$ of those games. (Note we need to divide by 2 since each game was counted twice). This means we have 96 games in total.

Answer: 96

Problem 16 Solution

The height of the parallelogram is 4 cm, so AD is equal to $48 \div 4 = 12$ cm. Since AE is 8 cm, ED must be $12 - 8 = 4$ cm long.

Hence, the area of the shaded triangle is $4 \times 4 \div 2 = 8$.

Answer: 8

Problem 17 Solution
First note that David types of 3 minutes, and since Charles started 2 minutes before David, Charles types for a total of

$$2 + 3 = 5$$

minutes. Since Charles types 12 words per minute faster than David, he can type

$$5 \times 12 = 60$$

more words than David in 5 minutes. Hence, if instead of Charles typing for 5 minutes and David typing for 3 minutes, David types for

$$5 + 3 = 8$$

minutes, he can type

$$780 - 60 = 720$$

words. Thus, David types

$$720 \div 8 = 90$$

words per minute. Since David actually types for 3 minutes, he types

$$3 \times 90 = 270$$

words in total. Lastly, we have Charles types the remaining

$$780 - 270 = 510$$

words.

Answer: 510

Problem 18 Solution

We can find equivalent fractions to $\dfrac{5}{11}$ by multiplying both the numerator and denominator by the same number. We can find some equivalent fractions until we get one that such that numerator and denominator add up to 80

$$\frac{5}{11} = \frac{10}{22} = \frac{15}{33} = \frac{20}{44} = \frac{25}{55} = \cdots$$

We can see $25 + 55 = 80$, so $m = 25$ and $n = 55$.

Answer: 55

Problem 19 Solution

The triangles that we would snip off the corners of the square are right isosceles triangles, where the length of the hypotenuse is the same as the length of the side of the octagon. Let's denote by x the side length. Then each of the legs of the right isosceles triangle has length $x \cdot \sqrt{2} \div 2$. Thus, the side of the square has length $x \cdot \sqrt{2} \div 2 + x + x \cdot \sqrt{2} \div 2 = x + x\sqrt{2}$, which is equal to 1 ft. So, the side length of the octagon is $x = \sqrt{2} - 1$, which means $B = 1$.

Answer: 1

Problem 20 Solution

If all 1600 ounces were from the first kind of orange juice, there would be

$$64\% \times 1600 = 0.64 \times 1600 = 1024$$

ounces of real orange juice. Debbie wants the juice to be 58% real orange juice, so she actually needs

$$58\% \times 1600 = 928$$

ounces of real orange juice. Hence she needs

$$1024 - 928 = 96$$

ounces less.

Each ounce of the first kind of orange juice contains 0.64 ounce of real orange juice, and each ounce of the second kind contains 0.48 ounce of real orange juice. Therefore each ounce of the second kind contains

$$0.64 - 0.48 = 0.16$$

ounce less real orange juice than the first kind. Debbie needs 96 ounces less, so she needs to replace

$$96 \div 0.16 = 600$$

ounces of the first kind with the second kind. Hence Debbie should use 600 ounces of the 48% real orange juice.

Answer: 600

2.6   ZIML March 2017 Division E

Below are the solutions from the Division E ZIML Competition held in March 2017.
The problems from the contest are available on p.51.

Problem 1 Solution
If your ticket number ends in 6, there are 9 tickets you can buy: 06, 16, 26, 36, 46, 56, 76, 86 and 96. Similarly, if your ticket starts with 6, there are another 9 tickets you can buy. Hence, there are a total of 18 tickets that you can buy.

Answer: 18

Problem 2 Solution
There is one vertical, one horizontal and 4 diagonal lines of symmetry as shown below.

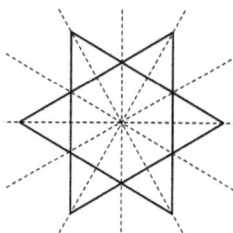

Answer: 6

Problem 3 Solution
Note that $1 \div \frac{1}{11} = 11$, $11 \div 1 = 11$, $121 \div 11 = 11$ and $1331 \div 121 = 11$. This tells us that to go from one number to the next we will need to multiply by 11, so the next number on the sequence is $1331 \times 11 = 14641$.

Answer: 14641

Problem 4 Solution

By looking at the diagram, we can see the longer side of each of the small rectangles is as long as 3 times the length of the short side, hence it will be $3 \times 2 = 6$ inches long. Then, the big rectangle has base 6 inches and height $2 + 6 + 2 = 10$ inches, so its perimeter is $2 \times (6 + 10) = 32$ inches.

Answer: 32

Problem 5 Solution

We know that the ratio of boys to girls is $7 : 4$, which is the same as $42 : 24$. Hence, if there are 42 girls participating in the program, there are 24 boys. This means there are $42 + 24 = 66$ participants in total.

Answer: 66

Problem 6 Solution

The smallest number that can be divided by both 6 and 10 is 30. All the numbers that can be divided by both 6 and 10 will be multiples of 30. The first multiple of 30 that is bigger than 100 is 120.

Answer: 120

Problem 7 Solution

We can pretend that we buy 6 more pencils instead of the pen and the total price we pay will be the same, but we will have now $6 + 1 = 7$ pencils. This means that each pencil costs $2.10 \div 7 = .30$ dollars, that is, 30 cents.

Answer: 30

Problem 8 Solution

The numbers you wrote are 103, 106, 109, 112, 115, 118 and

121. Their average is

$$\frac{103 + 106 + 109 + 112 + 115 + 118 + 121}{7} = 112.$$

Note that, since the numbers form an arithmetic sequence and you have an odd number of them, their average is the same as the number in the middle of the list.

Answer: 112

Problem 9 Solution
Lucy's triangle will have base $3 \times 2 = 6$ inches and height $4 \times 2 = 8$ inches. Hence, its area will be

$$\frac{6 \times 8}{2} = 24$$

square inches.

Answer: 24

Problem 10 Solution
In total, you and your friend have $20 + 25 = 45$ dollars. If you want to have twice as much money as your friend, that means that for every 1 dollar your friend has you will have 2 dollars, so, your friend will have $45 \div 3 = 15$ dollars and you will have $15 \times 2 = 30$ dollars. Since you already have 20 dollars, your friend needs to give you $30 - 20 = 10$ dollars.

Answer: 10

Problem 11 Solution
The base of the box, as well as the front and back sides of the box, each have area $3 \times 2 = 6$ square feet. The left and right sides of the box are squares with area $2 \times 2 = 4$ square feet. Hence, the total surface area of the open box is $6 + 6 + 6 + 4 + 4 = 26$

square feet.

Answer: 26

Problem 12 Solution

By the time the second train leaves the station, the first train has already advanced $2 \times 50 = 100$ miles. Every hour, the second train travels $60 - 50 = 10$ more miles than the first train. So, to make up for the extra 100 miles the first train has traveled so far, they must travel for $100 \div 10 = 10$ more hours.

Answer: 10

Problem 13 Solution

Each side of the big square will be $6 + 3 = 9$ inches long, this means that its area will be $9 \times 9 = 81$ square inches. Each of the triangles has area

$$\frac{1}{2} \times 3 \times 6 = 9$$

square inches, so the area of the small square is $81 - 4 \times 9 = 45$ square inches.

Answer: 45

Problem 14 Solution

The number that he keeps repeating has 10 digits in total. So, every 10th digit will be a 9, always followed by a 1. Since $144 \div 10 = 14$ with remainder 4, the last number Tom wrote will be the 4th number in the original 10-digit number, so 3.

Answer: 3

Problem 15 Solution

If a square has perimeter 20 each of its sides must measure $20 \div 4 = 5$. Hence the square has area $5 \times 5 = 25$, which is $25 - 21 = 4$

larger than the rectangle.

Answer: 4

Problem 16 Solution

We know the number of jellybeans is of the form $\overline{1a5}$. Notice that any multiple of 9 that ends in 5 is a multiple of $9 \times 5 = 45$. The multiples of 45 between 100 and 200 are 135 and 180. So, the bag had 135 jellybeans. Hence, each friend got

$$135 \div 9 = 15$$

jellybeans.

Answer: 15

Problem 17 Solution

If all sixty vehicles where motorcycles, we would have a total of

$$60 \times 2 = 120$$

wheels in the parking lot. We know there are actually 190 wheels in the lot,

$$190 - 120 = 70$$

more wheels than if we only have motorcycles. Switching a motorcycle for a car gives 2 extra wheels, therefore we must replace

$$70 \div 2 = 35$$

motorcycles with cars to get the correct number of wheels in the parking lot. Hence there are 35 cars and

$$60 - 35 = 25$$

motorcycles in the parking lot.

Answer: 35

Problem 18 Solution

Divide the figure into 4 identical triangles as in the diagram below.

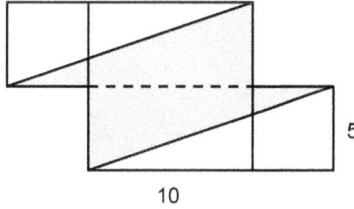

As 2 of the 4 triangles are shaded, we see the shaded region is

$$\frac{2}{4} = \frac{1}{2} = 50\%$$

of the full figure.

Answer: 50

Problem 19 Solution

In 1 minute, the hot tap will fill $\frac{96}{6} = 16$ gallons of water and the cold tap will fill $\frac{96}{4} = 24$ gallons of water. Since every minute 8 gallons go down the drain, we have $16 + 24 - 8 = 32$ gallons of water in the tub after 1 minute. Hence, it will take $96 \div 32 = 3$ minutes to fill in the tub.

Answer: 3

Problem 20 Solution

The numbers smaller than 40 that leave a remainder of 2 when you divide them by 5 are 2, 7, 12, 17, 22, 27, 32 and 37. We know it has to be an odd number, because it leaves a remainder of 1 when we divide it by 2, so it must be one of the numbers that end in 7. Finally, as it leaves no remainder when we divide it by

3, it must be a multiple of 3. The only multiple of 3 in our list that ends in 7 is 27, so that must be our number.

Answer: 27

2.7   ZIML April 2017 Division E

Below are the solutions from the Division E ZIML Competition held in April 2017.
The problems from the contest are available on p.57.

Problem 1 Solution
Consider dividing the rectangle into 8 identical triangles as shown below.

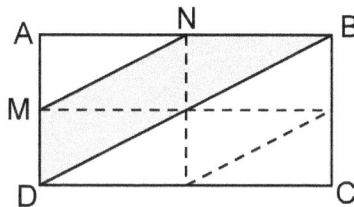

Therefore each of the triangles must have area

$$40 \div 8 = 5,$$

so the shaded region has area

$$3 \times 5 = 15.$$

Answer: 15

Problem 2 Solution
Notice that some numbers add up to 100. We can spot them easily because their units digits add up to 10 and their tens digits add up to 9.

$$\underline{\underline{54}} + \underline{67} + \underline{\underline{33}} + \underline{\underline{84}} + \underline{\underline{46}} + 64 + \underline{16}$$

This means we can add the numbers a lot quicker if we replace each of those pairs with a 100:

$$100 + 100 + 100 + 64 = 364.$$

Answer: 364

Problem 3 Solution

Since 40% of the students who play volleyball are boys, the other 60% are girls. Hence the number of girls who play volleyball is

$$60\% \times 180 = \frac{6}{10} \times 180 = 6 \times 18 = 108.$$

Answer: 108

Problem 4 Solution

We know that there are 24 candies in total. Thus, if they had the same number of candies, they would each have $24 \div 2 = 12$ candies. As Kyle needed to give 4 candies to Lisa for them to each have the same number, Kyle must have started with $12 + 4 = 16$ candies.

Answer: 16

Problem 5 Solution

If we subtract 2 from this number, the resulting number is a common multiple of 4 and 6. Since $\text{lcm}(4,6) = 12$, we are looking for the largest multiple of 12 that has two digits after adding 2. Now $96 = 8 \times 12$ is the largest two digit multiple of 12, and adding 2 to this number gives 98.

Answer: 98

Problem 6 Solution

During his walk, Jerry walks 3 blocks North and later 2 blocks South. This means at the end he is $3 - 2 = 1$ block North of his house. Similarly, he walks 4 blocks East and later 1 block West, so he is $4 - 1 = 3$ blocks East of his house. Therefore to get home, Jerry must walk 1 block South and 3 blocks West, a total of $1 + 3 = 4$ blocks.

Answer: 4

Problem 7 Solution

For Adam's square to have area 9, we see the side length must be 3 (as then $3 \times 3 = 9$ as needed). This is the height of Bob's triangle, so if Bob's triangle has area 3, we see the base must have length 2 (as then $\frac{1}{2} \times 3 \times 2 = 3$ as needed). Therefore Cynthia's square has a side length of 2, so area $2 \times 2 = 4$.

Answer: 4

Problem 8 Solution

To avoid over counting, label the people A, B, C, D, E, F, G, H. Note that A must shake hands with all the others, a total of 7 handshakes. B has then already shaken hands with A, so must shake hands with 6 new people. Similarly C must shake hands with 5 new people. This pattern continues, so in total there are

$$7 + 6 + 5 + 4 + 3 + 2 + 1 + 0 = 28$$

total handshakes.

Answer: 28

Problem 9 Solution

The first digit will be 2 and the second digit will be $2 \times 3 = 6$. Since we haven't used a 1 yet, we must have 1 ten and $1 \times 4 = 4$

---

ones. Hence the number in the sign is 2614.

Answer: 2614

Problem 10 Solution
If Jenny answered all questions correctly, she would have gotten a perfect score of
$$30 \times 5 = 150.$$

However, her resulting score was 114. For each incorrect or blank answer, Jenny loses the 5 points that the she would have gotten for the question and an additional point for the incorrect answer. Thus, her score decreases by a total of

$$5 + 1 = 6$$

points for each question missed. The total decrease in the score was
$$150 - 114 = 36$$

points, and thus the number of incorrectly answered or unanswered question was
$$36 \div 6 = 6.$$

Therefore Jenny answered

$$30 - 6 = 24$$

questions correctly.

Answer: 24

Problem 11 Solution
Labeling the vertices as in the figure below,

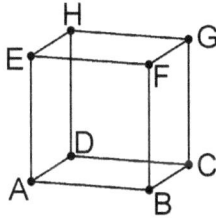

We can list the edges:

$$AB, BC, CD, DA, EF, FG, GH, HE, AE, BF, CG, DH.$$

Therefore there are a total of 12 edges.

Answer: 12

Problem 12 Solution

First we need to find the total distance Melisa traveled, which is

$$3 \times 50 + 2 \times 60 = 270 \text{ miles.}$$

We also need to find the total amount of time she traveled,

$$2 + 3 = 5 \text{ hours.}$$

We can then find Melisa's average speed, which is

$$270 \div 5 = 54 \text{ miles per hour.}$$

Answer: 54

Problem 13 Solution

First note each stick has a length of

$$100 \div 20 = 5 \text{ cm.}$$

Then we can divide the entire shape into squares with side length 5 cm:

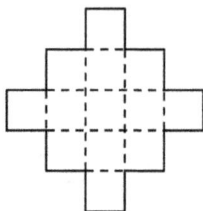

As there are $1 + 3 + 5 + 3 + 1 = 13$ such squares, each with area $5 \times 5 = 25$ square cm, the area of the shape is

$$13 \times 25 = 325 \text{ square cm.}$$

Answer: 325

Problem 14 Solution

Chris can paint $\dfrac{1}{12}$ of the house per hour, and Bill can paint $\dfrac{1}{8}$ of the house per hour. Together they can paint

$$\frac{1}{12} + \frac{1}{8} = \frac{2}{24} + \frac{3}{24} = \frac{5}{24}$$

of the house per hour. Therefore, it takes

$$\frac{24}{5} = 4.8$$

hours for them to paint the house together.

Answer: 4.8

Problem 15 Solution

Since we only care about the hundreds digit and the tens digit of the result, only the last 3 digits of both operands are important. So we simply multiply 606 and 707. Direct computation gets

$606 \times 707 = \ldots 8442$ (remember we only care about the hundreds and tens digits). Therefore $A = 4$, $B = 4$, so $A + B = 8$.

Answer: 8

Problem 16 Solution
If Will had typed by himself for 13 minutes, he would have typed

$$13 \times 40 = 520$$

words in total. That is, we would be still missing

$$600 - 520 = 80$$

words. Since Tim can type $50 - 40 = 10$ more words than Will every minute, if Tim types for

$$80 \div 10 = 8$$

minutes, he would make up for the missing words. Hence Tim typed for 8 minutes and Will typed for $13 - 8 = 5$ minutes.

Answer: 8

Problem 17 Solution
Originally the weight of the bananas plus basket was 32 pounds. After George ate half the bananas the weight of the remaining bananas plus basket was 17 pounds. Hence, the difference $32 - 17 = 15$ must be the weight of half the bananas. Subtracting this from 17 gives us the weight of the basket alone, which is $17 - 15 = 2$ pounds.

Answer: 2

Problem 18 Solution
The left and right sides of the parallelogram each have length 2 cm, so the remaining sides of the parallelogram have lengths that

add to 40 cm. As each triangle has a side length of 2 cm, there are

$$40 \div 2 = 20$$

triangles in total.

Answer: 20

Problem 19 Solution

Note that *SCIENCE* has 7 letters, so as $50 \div 7 = 7 \ R \ 1$, Paul could fit 7 copies of the word per line. Since $80 \div 7 = 11 \ R \ 3$, in 11 complete lines Paul can fit 77 copies of *SCIENCE*, so he will need one extra line for the remaining 3 copies. Hence Paul used 12 lines in total.

Answer: 12

Problem 20 Solution

We know that Susie's blocks each have a volume of 2 cubic units, so 5 of these blocks has a volume of 10 cubic units. As this must be equal to 2 of Frankie's blocks, one of Frankie's blocks has volume $10 \div 2 = 5$ cubic units. Since the full tower is built with 20 of Susie's blocks and 10 of Frankie's blocks, it must have volume

$$20 \times 2 + 10 \times 5 = 40 + 50 = 90.$$

Answer: 90

2.8   ZIML May 2017 Division E

Below are the solutions from the Division E ZIML Competition held in May 2017.
The problems from the contest are available on p.65.

Problem 1 Solution
Note we can break the whole figure into 28 small equilateral triangles.

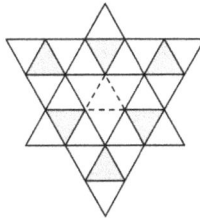

The shaded area is comprised of 6 of this triangles and has area 12, so each triangle has area $12 \div 6 = 2$. This means the whole figure has area $28 \times 2 = 56$

Answer: 56

Problem 2 Solution
Note $100 \div 11 = 9\ R\ 1$ and $1000 \div 11 = 90\ R\ 10$, so there are 90 multiples of 11 less than 1000 and 9 multiples of 11 less than 100. Hence there are $90 - 9 = 81$ multiples between 100 and 1000.

Answer: 81

Problem 3 Solution
If Quincey was missing no pencils, she would have $62 + 4 = 66$ pencils in total and each of the boxes would then have $66 \div 2 = 33$

pencils.

Answer: 33

Problem 4 Solution
If he chooses one of each cookie, he has only 1 way to do it. If he chooses two cookies of one kind and one cookie of a different kind, he can do it in $3 \times 2 = 6$ different ways. If he chooses all three cookies of the same kind, he can do it in 3 different ways. Thus, in total he has $1 + 6 + 3 = 10$ different ways of selecting the cookies.

Answer: 10

Problem 5 Solution
We can see 9 cubes, and there are 4 cubes hidden behind other cubes. Thus, $9 + 4 = 13$ cubes were used in total.

Answer: 13

Problem 6 Solution
On the same year the difference of the ages of Bob and his dad is 25. If we compare age of Bob this year and the age of his dad last year, the difference will be 24. We also know that the ratio of those ages is $1 : 3$. This means the ratio of Bob's current age and the difference of those ages is $1 : (3 - 1) = 1 : 2$. Therefore, Bob is $24 \div 2 = 12$ years old, and his dad is $12 + 25 = 37$ years old. The sum of their ages is $12 + 37 = 49$.

Answer: 49

Problem 7 Solution
Since the area of the big square is 144, each side of the big square must have length $\sqrt{144} = 12$. Note the side of the big square is equal to the sum of the length and width of the rectangle, and the side of the small square is equal to the difference between

the length and width of the rectangle. Since the length of the rectangle is 10, its width is $12 - 10 = 2$, and so the length of the side of the small square is $10 - 2 = 8$. Therefore, the area of the small square is $8 \times 8 = 64$.

Answer: 64

Problem 8 Solution

In total, you and your brother have $70 + 62 = 132$ marbles. Since now your brother has 3 times as many marbles as you, that means you have $\frac{1}{4}$ of the total number of marbles and your brother has the rest $\frac{3}{4}$. This means you actually have $\frac{1}{4} \times 132 = 33$, which tells us you lost $70 - 33 = 37$ marbles.

Answer: 37

Problem 9 Solution

Since 5 and 50 are counted by the same person, the number of people is a factor of $50 - 5 = 45$. 45 has factors $1, 3, 5, 9, 15, 45$, so there must have been 15 friends in total.

Answer: 15

Problem 10 Solution

The diagram can be broken down into 5 squares of the same size, each of area $180 \div 5 = 36$. Thus, the side of each of those small squares is 6. Since the perimeter of the figure uses 12 of those small sides, the perimeter of the figure is $12 \times 6 = 72$.

Answer: 72

Problem 11 Solution

Let's pretend Rita bought 4 less cans of Savory Shreds and 10 less cans of Prime Fillets. This means that she bought $38 - 4 - 10 = 24$ cans in total, and the same amount of each. So she must have bought $24 \div 3 = 8$ cans of Classic Pate.

Answer: 8

Problem 12 Solution
The code will have 4 characters in total. Since we are only choosing one letter, we can first decide in which of the 4 available spaces we want to put the letter. This can be done in 4 ways. Then we need to choose what letter to use and the digits to put in each of the remaining spaces. We only want to use consonants, so we have 21 letters available, and we only want odd digits, so we have 5 available. This means we can choose the code in $4 \times 21 \times 5 \times 5 \times 5 = 10500$ different ways.

Answer: 10500

Problem 13 Solution
The figure she wants to make looks like this:

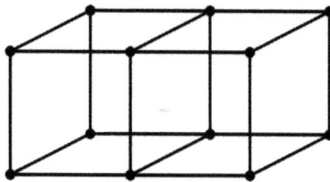

So she will need 20 toothpicks and 12 balls of clay. Hence the sum is $20 + 12 = 32$.

Answer: 32

Problem 14 Solution
The first pipe can empty $\frac{1}{6}$ of the pool per hour, and another pipe

can empty $\dfrac{1}{3}$ of the pool per hour. Together they can empty

$$\frac{1}{6} + \frac{1}{3} = \frac{1}{2}$$

of the pool per hour. Hence, it takes 2 hours for them to empty the pool together.

Answer: 2

Problem 15 Solution
If we fill both container with water, the first is 70% orange juice and the second is 80% orange juice. Since both containers have the same size, the percentage of orange juice after mixing is the average of the two containers. Therefore the large container contains

$$(70\% + 80\%) \div 2 = 75\%$$

orange juice.

Answer: 75

Problem 16 Solution
Note that the number we are looking for is actually the number of rectangles that are not squares. Since the grid we have is 3 by 3, the lengths and widths of our rectangles can be 1, 2 and 3. The dimensions of the rectangles that are not squares are 1×2, 1×3, and 2×3. We have 6 horizontal 1×2 rectangles, and 6 vertical 1×2 rectangles. We have 3 horizontal 1×3 rectangles and 3 vertical 1×3 rectangles. And we have 2 horizontal 2×3 and 2 vertical 2×3 rectangles. Therefore, we have $6 + 6 + 3 + 3 + 2 + 2 = 22$ rectangles that are not squares.

Answer: 22

Problem 17 Solution

Since each of the numbers on the list is the sum of the previous two, we can go backwards on the list by subtracting consecutive numbers. This way the fourth number on the list is the same as the difference of the last two, so $816 - 504 = 312$. Then the third number is $504 - 312 = 192$, the second number is $312 - 192 = 120$ and the first number (the smaller of Natalie's favorite numbers) is $192 - 120 = 72$.

Answer: 72

Problem 18 Solution
The location of Frank and Lacey is marked with an $\times$ below, with the ice cream shops also shown.

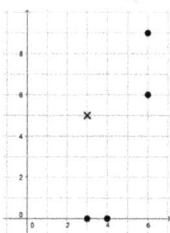

From this diagram we see that the ice cream shop at $(6,6)$ is the closest. Therefore the sum of the x and y-coordinates is $6 + 6 = 12$.

Answer: 12

Problem 19 Solution
Since every day he will add to his routine twice as many crunches as the day before, the fourth day he will add $2 \times 4 = 8$ for a total of $16 + 8 = 24$, and the fifth day he will add $2 \times 8 = 16$ for a total of $24 + 16 = 40$.

Answer: 40

Problem 20 Solution

Since the two friends are traveling in same direction, their relative speed is the difference of the two speeds, which is

$$65 - 55 = 10 \text{ miles per hour}$$

away from each other. After 2 hours of traveling, the difference in distance becomes,

$$10 \times 2 = 20 \text{ miles.}$$

Hence, they are 20 miles away from each other after 2 hours.

Answer: 20

2.9 ZIML June 2017 Division E

Below are the solutions from the Division E ZIML Competition
held in June 2017.
The problems from the contest are available on p.73.

Problem 1 Solution
Reordering we have

$$
\begin{aligned}
& 63 \times 25 \div 7 \times 2 \div 8 \div 5 \times 64 \div 9 \\
= \; & 63 \div 7 \div 9 \times 64 \div 8 \times 25 \div 5 \times 2 \\
= \; & 1 \times 8 \times 5 \times 2 \\
= \; & 80
\end{aligned}
$$

Answer: 80

Problem 2 Solution
Lemons are 4 for 1 dollar, so each lemon is

$$
1 \div 4 = \frac{1}{4}
$$

of a dollar. Similarly, each lime is

$$
1 \div 6 = \frac{1}{6}
$$

of a dollar. Julie buys 6 lemons, so Julie spends a total of

$$
6 \times \frac{1}{4} = \frac{6}{4} = \frac{3}{2} = 1.50
$$

dollars on lemons. She also buys 21 limes, so the amount of
money she spends on limes is

$$
21 \times \frac{1}{6} = \frac{21}{6} = \frac{7}{2} = 3.50
$$

dollars. Thus in total, Julie spends

$$1.50 + 3.50 = 5$$

dollars at the grocery store.

Answer: 5

Problem 3 Solution
There are always two sticky notes at the bottom left and top right. In between these sticky notes is a square made up of more sticky notes. In the first figure, there is no square, in the second it is a 1×1 square, in the third it is a 2×2 square, etc. Hence in the 10th figure it will be a 9×9 square. This square is made up of $9^2 = 81$ sticky notes, so in total there are $81 + 2 = 83$ sticky notes in the 10th figure.

Answer: 83

Problem 4 Solution
The square has area 144 square inches, so it must have a side length of 12 inches. As this side length is made up of 6 rectangles, we see that each rectangle has a width of $12 \div 6 = 2$ inches. As these rectangles all have length 12 inches, they have perimeter

$$2 \times 12 + 2 \times 2 = 24 + 4 = 28$$

inches.

Answer: 28

Problem 5 Solution
The positive factors of 36 are $1, 2, 3, 4, 6, 9, 12, 18, 36$. Therefore the sum is

$$1 + 2 + 3 + 4 + 6 + 9 + 12 + 18 + 36 = 91.$$

Answer: 91

Problem 6 Solution

There are $14 + 18 = 32$ liters of water in total. Since $32 \div 4 = 8$, we see if the blue bucket contains 8 liters of water and the red bucket contains $32 - 8 = 24$ liters of water then the red bucket will contain 3 times as much water as the blue bucket. Therefore we must pour $18 - 8 = 10$ liters of water from the blue bucket.

Answer: 10

Problem 7 Solution

If we plot the coordinates we get

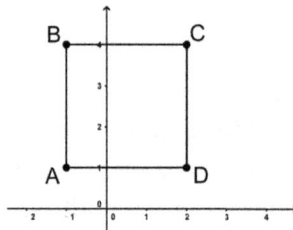

From this diagram it is easy to see that $ABCD$ is a square, and its side length is equal to $4 - 1 = 3$. Hence it has area $3^2 = 9$.

Answer: 9

Problem 8 Solution

If Troy used the dime, the two nickels and the two pennies he would get

$$10 + 2 \times 5 + 2 \times 1 = 22$$

cents in total, but he needs at least 23 cents to avoid getting all those pennies back.

By using one quarter he gets 25 cents, which is just a little bit more than the 23 cents he needs, so if he gives the cashier the $10 bill and one quarter, the cashier would have to give him back

$$10.25 - 6.23 = 4.02$$

dollars, that is, four $1 bills and two pennies. This is the least number of pennies Troy can receive, so our final answer is 2.

Answer: 2

Problem 9 Solution
If instead we have 99 unsupervised students with no teachers, the 99 tickets would cost a total of

$$99 \times 2 = 198$$

dollars. This is
$$218 - 198 = 20$$

dollars cheaper than the actual 99 tickets. If we remove one student ticket and add one teacher ticket, it costs $2 extra. Since $20 extra was paid over the price of 99 student tickets, there must be
$$20 \div 2 = 10$$

teachers.

Answer: 10

Problem 10 Solution
The base of the triangle is made up of 4 squares, so has length 4. Similarly, we see the height of the triangle is also 4. Hence the area of Molly's triangle is $\frac{1}{2} \times 4 \times 4 = 8$.

Answer: 8

Problem 11 Solution
We know Cameron, Mandy, and Taylor have $300 + $20 = $320 in total. Suppose instead that everyone had the same amount of money as Mandy. Then Taylor would have $20 extra and Cameron would have $50 more, so the friends would have

$$\$320 + \$20 + \$50 = \$390$$

in total. In this case they would all have $390 \div 3 = $130 dollars, so we know that Mandy must have originally had $130.

Answer: 130

Problem 12 Solution
First find the time that takes Terry to finish the race, which is

$$24 \div 6 = 4 \text{ hours.}$$

After 4 hours of running, the distance that Susan has run is

$$4 \times 4 = 16 \text{ miles.}$$

Since the race is 24 miles long, Susan still has

$$24 - 16 = 8 \text{ miles}$$

until she reaches the finish line.

Answer: 8

Problem 13 Solution
The square surrounding $\triangle ABC$ is 5×5 so has area 25. The square is made up of $\triangle ABC$ and three other triangles. The triangle on the left of $\triangle ABC$ has base 5 and height 1, the triangle above $\triangle ABC$ has base 4 and height 3, and the triangle below $\triangle ABC$ has base 5 and height 2. Therefore we can find the area of $\triangle ABC$ by subtracting:

$$25 - \frac{1}{2} \times 1 \times 5 - \frac{1}{2} \times 3 \times 4 - \frac{1}{2} \times 5 \times 2 = 25 - 2.5 - 6 - 5 = 11.5,$$

so $\triangle ABC$ has area of 11.5.

Answer: 11.5

Problem 14 Solution
Since 300 people were surveyed in total and we know 165 of them were female, a total of

$$300 - 165 = 135$$

males were surveyed. Similarly, a total of 207 people surveyed read the paper. Since 99 of them are female the other

$$207 - 99 = 108$$

were male. We can divide this by the total number of males

$$108 \div 135 = 0.8 = 80\%.$$

so see that 80% of the males surveyed read the newspaper.

Answer: 80

Problem 15 Solution
Including the next game, the pitcher will have pitched 9 total games after the next game. For a player to pitch an average of 6 innings, the sum of the number of innings pitched in the game is

$$6 \times 9 = 54.$$

He currently has

$$5 + 7 + 4 + 6 + 8 + 2 + 4 + 9 = 45$$

pitched. Therefore, he needs

$$54 - 45 = 9$$

innings in the next game.

Answer: 9

Problem 16 Solution

Since each niece gets 6 dollars if he distributes the bills to his nieces, the number of bills the uncle has must be a multiple of 6. Similarly, since each nephew gets 10 dollars, the number of bills he has must also be a multiple of 10. The only multiples of 10 less than 50 are $10, 20, 30, 40$, and only 30 is also a multiple of 6. Hence the uncle has 30 \$1 bills in total.

Answer: 30

Problem 17 Solution

Note a full rectangular box of this size will have length 10 inches, width 8 inches, and height 5 inches, so a volume of

$$10 \times 8 \times 5 = 400$$

cubic inches. To turn the described box into the table, we must remove a box with length $10 - 2 = 8$ inches, width 8 inches, and height $5 - 1 = 4$ inches, so a volume of

$$8 \times 8 \times 4 = 256$$

cubic inches. Hence Bob's model table has a volume of $400 - 256 = 144$ cubic inches.

Answer: 144

Problem 18 Solution

Stan can load $\dfrac{1}{40}$ of the truck per minute. When we work together, we can load $\dfrac{1}{15}$ of the truck per minute. Then, we can find the amount of work that I can do per minute

$$\frac{1}{15} - \frac{1}{40} = \frac{1}{24}.$$

Since I can load $\dfrac{1}{24}$ of the truck per minute, it would take me 24 minutes to finish it alone.

Answer: 24

Problem 19 Solution
Let's take a look at the numbers in the list from 100 to 199. They all have a 1 as the first digit. There are only 10 numbers that have 0 in the tens digit: 100, 101, 102, ..., 108 and 109, but 100 has two 0s, so we have 9 numbers with exactly one 0 so far. The rest of the numbers that will have 0 will have it in the ones digit: 110, 120, 130, ..., 190. So, from 100 to 199 we have $9 + 9 = 18$ numbers that have exactly one 0.

Note the same counting works for the 200s, the 300s, the 400s, etc. Since we have 9 possible digits we can choose for the hundreds digit, we will have in total $18 \times 9 = 162$ numbers with exactly one 0.

Answer: 162

Problem 20 Solution
The left and right sides each have length 2 cm, so the remaining sides of the rectangle have length $144 - 2 - 2 = 140$ cm. Therefore, the length of the top of the long parallelogram is

$$140 \div 2 = 70 \text{ cm}.$$

As the top of each square is 2 cm, there are

$$70 \div 2 = 35$$

total squares.

Answer: 35

3. Appendix

3.1 Division E Topics Covered

Note: Setting up and solving equations is not necessary for the problems in Division E (there are rare exceptions among the earlier monthly contests though). Students are allowed to use equations to solve the questions, but the questions are designed to be solved without using equations or systems of equations.

Word Problems

- Calculations and Arithmetic: Adding, subtracting, multiplying, and dividing whole numbers, fractions, and decimals
- Ratios and Proportions: Using ratios to find parts of a whole, Calculating missing information from proportional relationships, etc.
- Percents: Calculating percent increases and decreases, Relationship between percents and ratios, Using percents in mixture problems (e.g. 40% water and 60% oil)
- Problem Solving Methods: Chicken and Rabbit method, Using ratios when given sums or differences
- Motion Problems using (Speed)x(Time)=(Distance), Average Speed, Applying proportions to motion problems

- Work using (Rate)x(Time)=(Work Done), Average Rate of Work, Applying proportions to work problems

Geometry

- Areas and Perimeters of Basic Shapes such as triangles, rectangles, parallelograms, trapezoids, and circles
- Symmetry of Polygons
- Similar Triangles: Equal Angles, Sides are in a common ratio
- Geometric Reasoning with Areas: Congruent shapes have the same area, Dividing a shape and rearranging areas to find patterns, etc.
- Volumes and Surface Areas of Basic Solids such as cubes and rectangular prisms (boxes)

Number Sense

- Place Values: Ones/units digit, tens digit, hundreds digit, etc.
- Fundamental Definitions: Quotients and Remainders, Prime numbers, Factors (Divisors), Multiples, Perfect squares, Perfect cubes, etc.
- Least common multiple (LCM), Greatest common factor or divisor (GCF or GCD)
- Sum and Product Rules for Counting
- Sequences: Arithmetic and Geometric Sequences, Sum of elements in an arithmetic sequence, Finding patterns for general sequences
- Probability: Gives the chance of something happening, Ratio of outcomes
- Basic Statistics: Mean (Average), Median, Mode for lists, Interpreting data from graphs, bar charts, tables, etc.

3.2 Glossary of Common Math Terms

Acute Angle An angle less than $90°$.

Altitude of a Triangle A line segment connecting a vertex of a triangle to the opposite side forming a right angle. Also called the height of a triangle.

Angle A figure formed by two rays sharing a common vertex. Often measured in degrees.

Arc The curve of a circle connecting two points.

Area The amount of space a region takes up. Often denoted using square brackets: area of $\triangle ABC = [ABC]$.

Arithmetic Sequence A sequence where the difference between one term and the next is constant.

Average See Mean.

Base of a Triangle One side of a triangle, often used when the altitude is drawn from the opposite side to this base.

Chord A line segment connecting two points on the outside of a circle.

Circle A round shape consisting of points that all have the same distance (called the radius) from the center of the circle.

Circumference The perimeter of a circle.

Composite Number A number that is not prime.

Congruent Two shapes or figures that are exactly the same.

Cube A solid figure formed by 6 congruent squares that all meet at right angles.

Deck of Cards A standard deck of cards has 52 cards. There are 4 suits (clubs, diamonds, hearts, and spades) with each suit having cards of 13 ranks (A (ace), $2, 3, \ldots, 10$, J (jack), Q (queen), and K (king)).

Denominator The bottom number in a fraction.

Diagonal A line segment connecting two vertices of a shape or solid that is not an edge of the shape or solid.

Diameter A chord passing through the center of a circle. The diameter has length that is twice the radius.

Die or Dice A standard die (plural is dice) has 6 sides. Each of the 6 sides has the same chance when the die is rolled.

Digit One of $0, 1, 2, \ldots, 9$ used when writing a number.

Distinguishable Objects Objects that are different.

Divisible A number is divisible by another number if there is no remainder when the first number is divided by the second. For example, 35 is divisible by 7.

Divisor A number that evenly divides another number. For example, 6 is a divisor of 48. Also called a factor.

Edge A line segment connecting two vertices on the outside of a shape or solid.

Equally Likely Having the same chance of occurring.

Equiangular Polygon A shape with all equal angles.

Equilateral Polygon A shape with all equal sides.

Equilateral Triangle A regular triangle, one with three equal sides and three equal angles.

Even Number A number divisible by 2.

Exponent The number another number is raised to for powers. For example, in a to the power of b (a^b), the exponent is b.

Face The shape or polygon on the outside of a solid region.

Factor of a Number A number that evenly divides another number. For example, 6 is a factor of 48. Also called a divisor.

Factorial The symbol ! where $n! = n \times (n-1) \times (n-2) \cdots \times 1$.

Fraction An expression of a quotient. For example, $\frac{1}{2}$ or $\frac{9}{7}$.

Geometric Sequence A sequence where the ratio between one term and the next is constant.

Greatest Common Divisor (GCD) The largest number that is a divisor/factor of two or more numbers.

Greatest Common Factor (GCF) See Greatest Common Divisor.

Indistinguishable Objects Objects that are the same.

Isosceles Triangle A triangle with two equal sides and two equal angles.

Least Common Multiple (LCM) The smallest number that is a multiple of two or more numbers.

Mean The sum of the numbers in a list divided by the how many numbers occur in the list. Also called the average.

Median The number in the middle of a list when the list is arranged in increasing order.

Midpoint The point in the middle of a line segment.

Mode The number or numbers occurring most often in a list of numbers.

Multiple A number that is an integer times another number. For example, 72 is a multiple of 8.

Numerator The top number in a fraction.

Obtuse Angle An angle between $90°$ and $180°$.

Odd Number A number not divisible by 2.

Parallel Lines Lines that do not intersect.

Perfect Cube A number that is another number cubed. For example, $64 = 4^3$ is a perfect cube.

Perfect Square A number that is another number squared. For example, $64 = 8^2$ is a perfect square.

Perimeter The length/distance around the outside of a shape.

Pi (π) A number used often in geometry. $\pi = 3.1415926\ldots \approx 3.14 \approx \frac{22}{7}$.

Polygon A shape formed by connected line segments.

Prime Factorization The expression of a number as the product of all its prime factors. For example, 24 has prime factorization $2 \times 2 \times 2 \times 3 = 2^3 \times 3$.

Prime Number A number whose only factors are one and itself.

Proportional Ratios Ratios that have equal values when expressed in fraction form. For example, $2 : 3$ is proportional to $8 : 12$.

Quadrilateral A shape with four sides.

Quotient The integer quantity when dividing one number by another. For example, the quotient of $38 \div 5$ is 7 as $38 = 7 \times 5 + 3$.

Radius of a Circle The distance from the center of the circle to any point on the outside of the circle.

Randomly Chosen for a group of objects. Unless specified, the chance of choosing each object is the same as any other object.

Rank of a Card See Deck of Cards.

Ratio A relation depicting the relation between two quantities. For example $2 : 3$ or $\frac{2}{3}$ denotes that for every 3 of the second quantity there are 2 of the first quantity.

Rectangle A quadrilateral with four right angles (an equiangular quadrilateral).

Regular Polygon A polygon with all equal sides and all equal angles (equilateral and equiangular).

Remainder The quantity left over when one integer is divided by another. For example, the remainder of $38 \div 5$ is 3 as $38 = 7 \times 5 + 3$.

Rhombus A quadrilateral with four equal sides (an equilateral quadrilateral).

Right Angle A $90°$ angle.

Right Triangle A triangle containing a right angle.

Scalene Triangle A triangle with three unequal sides and three unequal angles.

Sequence An ordered list of numbers.

Similar Shapes or solids that have the same angles and sides that share a common ratio.

Square A shape with four equal sides and four equal angles (a regular quadrilateral).

Suit of a Card See Deck of Cards.

Surface Area The total area of all the faces of a solid.

Trapezoid A quadrilateral with one pair of parallel sides.

Triangle A shape with three sides.

Vertex The intersection of line segments, especially the intersection of sides or edges in a shape or solid.

Volume The amount of space a solid region takes up.

With Replacement When choosing objects with replacement, a chosen object is returned to the others allowing it to be chosen more than once.

3.3 ZIML Answers

ZIML October 2016 Division E

Problem 1:	150	Problem 11:	8
Problem 2:	728	Problem 12:	360
Problem 3:	25	Problem 13:	24
Problem 4:	15	Problem 14:	20
Problem 5:	824	Problem 15:	26
Problem 6:	5	Problem 16:	1
Problem 7:	4	Problem 17:	32
Problem 8:	-72	Problem 18:	15
Problem 9:	5	Problem 19:	25
Problem 10:	28	Problem 20:	6

ZIML November 2016 Division E

Problem 1:	15	Problem 11:	20
Problem 2:	37.5	Problem 12:	3
Problem 3:	16	Problem 13:	22
Problem 4:	11	Problem 14:	7
Problem 5:	53	Problem 15:	6750
Problem 6:	50	Problem 16:	36
Problem 7:	5	Problem 17:	55
Problem 8:	7	Problem 18:	21
Problem 9:	45	Problem 19:	13
Problem 10:	6	Problem 20:	36

ZIML December 2016 Division E

Problem 1:	2100	Problem 11:	2
Problem 2:	1348	Problem 12:	7
Problem 3:	5	Problem 13:	55
Problem 4:	105	Problem 14:	216
Problem 5:	24	Problem 15:	30
Problem 6:	900	Problem 16:	8
Problem 7:	30	Problem 17:	92
Problem 8:	3	Problem 18:	5
Problem 9:	100	Problem 19:	130
Problem 10:	100	Problem 20:	4

ZIML January 2017 Division E

Problem 1:	896	Problem 11:	96
Problem 2:	90	Problem 12:	42.5
Problem 3:	4	Problem 13:	4
Problem 4:	48	Problem 14:	40
Problem 5:	1249	Problem 15:	405
Problem 6:	64	Problem 16:	5
Problem 7:	2	Problem 17:	12
Problem 8:	30	Problem 18:	124
Problem 9:	35	Problem 19:	14
Problem 10:	25	Problem 20:	68

ZIML February 2017 Division E

Problem 1:	13	Problem 11:	8
Problem 2:	11	Problem 12:	10
Problem 3:	31	Problem 13:	65
Problem 4:	16	Problem 14:	75
Problem 5:	7	Problem 15:	96
Problem 6:	4	Problem 16:	8
Problem 7:	2232	Problem 17:	510
Problem 8:	5	Problem 18:	55
Problem 9:	21	Problem 19:	1
Problem 10:	89	Problem 20:	600

ZIML March 2017 Division E

Problem 1: 18

Problem 2: 6

Problem 3: 14641

Problem 4: 32

Problem 5: 66

Problem 6: 120

Problem 7: 30

Problem 8: 112

Problem 9: 24

Problem 10: 10

Problem 11: 26

Problem 12: 10

Problem 13: 45

Problem 14: 3

Problem 15: 4

Problem 16: 15

Problem 17: 35

Problem 18: 50

Problem 19: 3

Problem 20: 27

ZIML April 2017 Division E

Problem 1:	15	Problem 11:	12
Problem 2:	364	Problem 12:	54
Problem 3:	108	Problem 13:	325
Problem 4:	16	Problem 14:	4.8
Problem 5:	98	Problem 15:	8
Problem 6:	4	Problem 16:	8
Problem 7:	4	Problem 17:	2
Problem 8:	28	Problem 18:	20
Problem 9:	2614	Problem 19:	12
Problem 10:	24	Problem 20:	90

ZIML May 2017 Division E

Problem 1: 56

Problem 2: 81

Problem 3: 33

Problem 4: 10

Problem 5: 13

Problem 6: 49

Problem 7: 64

Problem 8: 37

Problem 9: 15

Problem 10: 72

Problem 11: 8

Problem 12: 10500

Problem 13: 32

Problem 14: 2

Problem 15: 75

Problem 16: 22

Problem 17: 72

Problem 18: 12

Problem 19: 40

Problem 20: 20

ZIML June 2017 Division E

Problem 1:	80	Problem 11:	130
Problem 2:	5	Problem 12:	8
Problem 3:	83	Problem 13:	11.5
Problem 4:	28	Problem 14:	80
Problem 5:	91	Problem 15:	9
Problem 6:	10	Problem 16:	30
Problem 7:	9	Problem 17:	144
Problem 8:	2	Problem 18:	24
Problem 9:	10	Problem 19:	162
Problem 10:	8	Problem 20:	35

www.ingramcontent.com/pod-product-compliance
Lightning Source LLC
Chambersburg PA
CBHW050124210326
41519CB00015BA/4100